Success Factors for Minorities in Engineering

T0174220

This book aims to isolate specific success factors for underrepresented minorities in undergraduate engineering programs. Based on a three-phase study spearheaded by the National Action Council for Minorities in Engineering, the findings include evidence that hands-on exposure to problem-based courses, research, and especially internships are powerful catalysts for engineering success, and that both college adjustment and academic skills matter, in varying degrees, to minority success. By encompassing an unusually large number and range of programs, this research adds to the evidence base for the importance of hands-on exposure to the work of engineering.

Jacqueline Fleming is a psychologist and independent researcher based in Pearland, Texas.

Irving Pressley McPhail is Founder and Chief Strategy Officer at The McPhail Group LLC.

Routledge Research in Higher Education

For more information about this series, please visit: www.routledge.com/
Routledge-Research-in-Higher-Education/book-series/RRHE

Success Factors for Minorities in Engineering

Jacqueline Fleming
and
Irving Pressley McPhail

Routledge
Taylor & Francis Group

NEW YORK AND LONDON

First published 2019
by Routledge
52 Vanderbilt Avenue, New York, NY 10017

and by Routledge
2 Park Square, Milton Park, Abingdon, Oxon, OX14 4RN

First issued in paperback 2020

Routledge is an imprint of the Taylor & Francis Group, an informa business

Library of Congress Cataloguing in Publication Data
A catalog record for this title has been requested

ISBN 13: 978-0-367-66079-6 (pbk)
ISBN 13: 978-1-138-38550-4 (hbk)

Typeset in Sabon
by Apex CoVantage, LLC

Contents

Preface

This volume is the result of a long-awaited collaboration between two people who have admired each other's work from afar until now. Irving Pressley McPhail, then President and CEO of the National Action Council for Minorities in Engineering, felt strongly that the organization's efforts would benefit from a more precise knowledge of what best facilitates the successful retention and graduation of underrepresented minorities in undergraduate engineering education. Several smaller studies had gotten the ball rolling, but a more comprehensive view was needed both to inform the scholarships that NACME provided to 31 Block Grant institutions with strong Minority Engineering Programs (MEPs) as well as to inform the programmatic efforts at NACME's 48 partner institutions, inclusive of the Block Grant institutions. Having watched Jacqueline Fleming's work over the years, a fortuitous meeting launched their combined effort.

Jacqueline Fleming, a research psychologist, had made a career of evaluating the education environments and developmental progress of minority students but developed a soft spot for engineering students from evaluating programs for them at various engineering schools including the City College of New York, California State University at Northridge, and Xavier University of New Orleans. Convinced that they were a special breed of student, she quickly warmed to the idea of working with them if daunted by the prospect of doing so at 31 potential universities.

Fortunately, the National Science Foundation also warmed to the idea of a broad study that concentrated exclusively on engineering students with an emphasis on underrepresented minority students. It is an understatement to say that we are grateful for their support, and extend our thanks to our two program officers, Donna Riley and Elliot Douglas.

As with any project that spans years, this one has many people to acknowledge. It would not have been possible without the management as well as support from Christopher Smith, then Director of Research for NACME. As the managing liaison between Fleming, NACME and 31 engineering schools, there is a special place in heaven for someone possessed of such calm at the intersection of hysteria and bureaucratic entanglements. At each of the universities, the designated staff members assisted the project with great competence, valiantly adding to their already overburdened work load.

And then there is the long list of intelligent and industrious others whose praises must be sung: the heroic efficiency of technical consultant Raijanel Crockem of Texas Southern University, the top-notch research assistance of Brenda Gamez-Patience and Shoquiong Lin of Texas A&M University; Shannon Thomas for her help with the nightmare of University Research Review Boards; Darlene Schluter for unfailing energy in the face of the tedious demands of research, and Dr. Marguerite Bonous-Hammarth for her review of our early proposals. Our study consultants, Drs. Henry Frierson of Florida A&M and Charles Watkins of the City College of New York, provided not only expert knowledge and advice but also much needed support. We thank you.

From Irv: to the power of family—Mauise Vinson McPhail, Pressley Samuel McPhail, Christine Johnson McPhail, Kamilah McPhail McKissick, Juan Detausia McKissick, Connor Pressley McKissick, and Chase Joseph McKissick—for your love, support, courage, and encouragement.

From Jackie: to the power of living with Leslie Calvin Hill Jr.

Jacqueline Fleming
Pearland, Texas
JacquelineFleming@yahoo.com
and
Irving Pressley McPhail
Amawalk, New York
imcphail@themcphailgroup.com

Figures

Tables

Appendices

1 The National Action Council for Minorities in Engineering

Its History and Mission

Since its inception in 1974, National Action Council for Minorities in Engineering, Inc. (NACME) has remained true to its mission: to ensure American resilience in a flat world by leading and supporting the national effort to expand U.S. capability through increasing the number of successful African American, Latino, and Native American/Alaska Native women and men in science, technology, engineering, and mathematics (STEM) education and careers. The founding mandate of NACME was to conduct ongoing research, to identify the impediments limiting access to careers in engineering for underrepresented minorities (URMs), and to implement programs to achieve a technical workforce truly reflective of the American population (see also Pierre, 2015).

Scholarships became the central strategic thrust in the 1970s, and, today, NACME is the largest private provider of scholarships for URMs in engineering education. With funding from corporate and individual donors, NACME has supported more than 24,000 students with more than $142 million in scholarships and other support, and currently has 1,300 scholars at 31 Block Grant institutions across the nation. Through its administration of the Alfred P. Sloan Foundation's Minority Ph.D. Program, 1,044 Ph.D. degrees in STEM disciplines have been produced. In addition, the Sloan Indigenous Graduate Partnership (SIGP) has graduated 27 Ph.D. degrees and 103 Master's degrees in STEM disciplines.

The NACME strategy over the past 44 years has also embraced a strong commitment to research and program evaluation, engineering public policy, and pre-engineering programs. NACME remains the most authoritative source of data and engineering public policy recommendations focused on moving the needle in URM participation in engineering education and careers (NACME, 2008, 2011, 2013). NACME also partnered with the National Academy Foundation and Project Lead the Way in 2006–2015 to open Academies of Engineering (AOEs) across the nation. The AOEs are urban-centered, open enrollment, high school-level engineering academies that provide all students with a strong science and mathematics education so that they will be college-ready for engineering education. In the AOE paradigm, NACME board companies and other industry partners provided internship and mentoring experiences for students designed to give them active, real-world experiences in engineering and innovation (McPhail, 2010, 2011).

University Programs

It is instructive to trace the evolution of the NACME scholarship strategy as context for the present study (NACME, 2010). Beginning in the 1970s, NACME, through the National Fund for Minority Engineering Students (NFMES), introduced the Incentive Grants Program (IGP). IGP provided universities with large grants with the expectation that they would provide additional university funding targeted at recruiting, enrolling, educating, retaining, and graduating increasing numbers of underrepresented minority engineering students. The program was initially successful, increasing the number of URM freshman in engineering study from 2,249 in 1973 to 1,116 in 1981. However, this rapid growth stalled through the 1980s, before bouncing back to 11,754 by the end of the decade.

NACME's research department identified two factors accounting for these results: (1) the pool of URMs graduating from high school ready to enter undergraduate engineering education had remained stagnant during the late 1970s, marking 1980–1981 as a saturation point in the minority engineering recruitment effort; and (2) the growth in enrollment in engineering study was accompanied by a significant increase in student attrition. The consequence of this research was a re-direction of NACME resources to the establishment of Minority Engineering Programs (MEPs) on the university campuses. The MEPs acknowledged both the intensity of the undergraduate engineering education curriculum and the less than optimal conditions for learning at Predominantly White Institutions (PWIs) for URMs. The MEPs aimed to strengthen the academic and survival skills of URM engineering students.

With seed funding and technical assistance, NACME led the development of 11 MEPs in 1980–1981. NACME's research efforts led to improvements in the MEP model, and, in collaboration with the National Association of Minority Engineering Program Administrators (NAMEPA), NACME published a best practices handbook on how to start and operate effective MEPs (see Landis, 1985). Additional print resources focused on academic success, financial aid, and career guidance were developed by NACME to optimize the national MEP effort, allowing NACME to touch the lives of thousands of URMs in engineering programs.

By 1980, the IGP was spending approximately $3 million annually and supporting about 12% of all URMs enrolled in undergraduate engineering education. However, the shifting landscape of student learning outcomes and college affordability resulted in NACME once again shifting the focus of the IGP from outreach and recruitment to scholarship support and best practices in retention to graduation.

The current NACME Block Grant Scholarship Program reduced the number of universities receiving grants from 140 to 31, focusing on those institutions with the best retention to graduation records. The minimum scholarship amount for individual students was increased, while the number of scholarships awarded was reduced. Universities that wish to participate in the NACME Block Grant Scholarship Program must document metrics that manifest the institutional commitment to recruiting, enrolling, educating, retaining, and graduating African American, Latino, and Native American/Alaska Native women and men in undergraduate engineering education. The program provides scholarship support in the form of a lump sum grant to partner institutions that enroll students from three sources—first-year students identified by NACME or the Partner Universities, transfer students from community colleges, and currently enrolled students who have completed at least one year of engineering study.

The NACME scholarship and university strategy has proven to be an eminently successful paradigm for undergraduate minority engineering education for over four decades. NACME Scholars maintain an unprecedented 79.1% six-year retention to graduation rate, compared to 39.3% for non-NACME Scholar URMs, and 60.3% for non-minority students (NACME, 2016). NACME Scholars earn an average GPA of 3.35/4.00 scale. Each year, the NACME Partner Institutions account for more than 30% of the total number of bachelor's degrees in engineering awarded to African American, Latino, and Native American/Alaska Native students. One-third of NACME Scholars are first-generation college students. Poised to be leaders in the engineering workforce, 53% of NACME Scholars who participated in internship and co-op experiences in 2015 at 113 different companies said they would go back to work at the company based on their experience.

The "New" American Dilemma

In 2008, NACME released a landmark report titled *The "New" American Dilemma: A Data-Based Look at Diversity* (NACME, 2008). The term "American dilemma" was originally coined by Swedish social scientist Gunnar Myrdal. Myrdal's two-volume work by that title, commissioned by the Carnegie Corporation, examined the status of race relations in the United States (Myrdal, 1944). Published near the end of World War II in 1944, Myrdal found that despite a strong ethos of equality, African Americans were subjected to searing inequality. On any of the critical dimensions of survivability—health care, schools, jobs, housing, social facilities, etc.—the 10% of Americans of African descent at this time were far worse off than their Non-Latino White compatriots. Myrdal's study laid bare the contradictions between the rhetoric of equality on the one hand and the reality of inequality on the other.

In the forward to the 2008 report, Dr. John Brooks Slaughter, President and CEO at NACME, aligned the dilemma described by Myrdal as a pressing social problem with the present dilemma of increasing the

representation of African Americans, Latinos, and Native American/
Alaska Natives in engineering. He argued:

> Our purpose is to send a clear and unambiguous message that must
> be understood and acted upon if this nation is to retain its position
> of leadership in STEM and keep its competitive edge in the global
> marketplace of ideas and products.
>
> That message is this: The solution to America's competitiveness
> problem is to activate the hidden workforce of young men and
> women who have traditionally been underrepresented in STEM
> careers—African Americans, American Indians, and Latinos.
>
> (p. 3)

Resolving this dilemma is a matter of increasing national import, as
we have become a much more diverse and "flat" world. Engineers are
the visionaries of the future. A diverse engineering workforce is key to
maintaining our competitive edge in an increasingly global economy.
Yet the founding vision of NACME of an engineering workforce that
looks like America has remained "a dream deferred"[1] over the past four
decades.

To more fully grasp the dimensions of this challenge, consider the
major findings from the *2011 NACME Data Book: A Comprehensive
Analysis of the "New" American Dilemma* (NACME, 2011):

The U.S. Population Is Becoming More Diverse

- Between 2010 and 2050, the relative percentage of the U.S. population that is Non-Latino White is expected to decline from 65% in
  2010 to 46% in 2050. By then, Latinos will account for 30% of the
  U.S. population and Asians for 8%.
- Already, 43% of school-aged children (aged 5–17) are African American, Latino, Native American or Asian/Pacific Islander Americans.
- Underrepresented minorities (URMs) account for 34% of the
  18–24-year-old U.S. population.
- Forty percent of 18–24-year-olds are from ethnic minority groups;
  34% are URMs.

The Pipeline to Engineering Is Far from Full

- Twenty percent of Latinos (males) and 14% of Latinas drop out of
  high school.
- Fewer than 8% of Latino, African American, and Native American
  high school seniors take calculus vs. 15% of Non-Latino Whites and
  30% of Asian American seniors.

- The gender gap in high school preparation in advanced science and mathematics has disappeared; high school senior females and males are equally likely to take calculus, analysis/pre-calculus, chemistry, and physics.
- Underrepresented minority youth are less likely than Non-Latino White and Asian students to complete a "rigorous" high school curriculum.

  - Four years English.
  - Three years social studies.
  - Four years mathematics, including pre-calculus or higher.
  - Three years of science, including biology.

Enrollment Rates Are Increasing

- Enrollment rates have increased for all groups over the past 30 years.
- Community college enrollment has increased and has become a significant starting point for many students' postsecondary education.
- Retention to graduation in engineering continues to be low with important variations across racial/ethnic categories.

Engineering Schools Are Not Tapping the Diverse U.S. Talent Pool

- URMs earned just 13% of all engineering bachelor's degrees in 2009.
- Women earned just 18% of engineering bachelor's degrees in 2009.
- There were 7,915 doctoral degrees awarded in engineering in 2009, of which 57% were awarded to temporary residents.
- Eleven percent of engineering doctoral degrees were awarded to U.S. women.
- URMs earned just 4% of the nation's doctoral degrees in engineering in 2009, representing a total of 311 potential new faculty members for the more than 300 colleges of engineering in the U.S.

U.S. Engineering Is Robust Even in Difficult Times, Yet Does Not "Look like America"

- Bureau of Labor Statistics projects a need for 178,300 more engineers in the next decade with the fastest growth in biomedical, civil, environmental, industrial, and petroleum engineering.
- New engineering bachelor's degree graduates continue to earn very high starting salaries (average $59, 435 in 2011), and the engineering unemployment rate of 6.9% is lower than that for all workers (9.3%).

- Nearly flat trajectories associated with Latino and African American representation on engineering faculties suggests that minority engineering students will continue to lack professors that "look like" them for many years to come.
- Salaries of men and women do not differ substantially within each of the ethnic/racial categories.
- Salaries of African American and Latino engineers are not on par with those of Non-Latino White engineers.
- Asian American engineers' salaries are higher, on average, than those of other engineers, which may be due to specific industry or geographic factors.

Against the backdrop of *The "New" American Dilemma*, this book looks at performance and retention to graduation rates of minority students in NACME Block Grant Institutions. The authors believe that is important to identify what works and why in undergraduate minority engineering education, as opposed to continuing to document failure. Achieving NACME's vision of an engineering workforce that looks like America requires that more be done to exponentially increase the number of URMs successfully completing the bachelor's degree in engineering. It is our hope that this book will contribute to the knowledge base in undergraduate minority engineering education, and, at the same time, encourage more URMs to pursue engineering degrees.

Note

1 *What Happens to a Dream Deferred* is one of a number of poems by Langston Hughes that speaks to the condition of African American people in the United States. The poem explores the dreams and aspirations of a people, and ponders the consequences that might arise if those dreams and aspirations are not realized. In the context of the present argument, the fact that the domestic engineering workforce does not look like America is *a dream deferred*. The consequences of the underrepresentation of African American, Native American, and Latino women and men in engineering education and careers pose serious threats to the United States' stature in STEM on a global scale.

References

Landis, R. B. (1985). *Improving the retention and graduation of minorities in engineering*. New York: National Action Council for Minorities in Engineering.

McPhail, I. P. (2010, Winter). Mentoring is key to building a diverse workforce. *Winds of Change, 25*(1), 80.

McPhail, I. P. (2011, November). Last word: Industry can help us diversify: Budget woes make the private sector a crucial player in STEM. *ASEE Prism, 21*(3), 80.

Myrdal, G. (1944). *An American dilemma: The Negro problem and modern democracy* (2 Vols.). New York: Harper & Brothers.

NACME (National Action Council for Minorities in Engineering). (2008). *Confronting the "new" American dilemma: A data-based look at diversity*. White Plains, NY: National Action Council for Minorities in Engineering.

NACME (National Action Council for Minorities in Engineering). (2010). *NACME alumni today: 2011*. White Plains, NY: National Action Council for Minorities in Engineering.

NACME (National Action Council for Minorities in Engineering). (2011). *2011 NACME data book: A comprehensive analysis of the "new" American dilemma:* White Plains, NY: National Action Council for Minorities in Engineering.

NACME (National Action Council for Minorities in Engineering). (2013). *2013 NACME data book: A comprehensive analysis of the "new" American dilemma*. White Plains, NY: National Action Council for Minorities in Engineering.

NACME (National Action Council for Minorities in Engineering). (2016). *About NACME: Engineering a workforce that looks like America*. White Plains, NY: NACME.

Pierre, P. A. (2015). A brief history of the collaborative minority engineering effort: A personal account. In J. B. Slaughter, Y. Tao, & W. Pearson, Jr. (Eds.), *Changing the face of engineering: The African American experience* (pp. 13–35). Baltimore, MD: Johns Hopkins University Press.

2 Minorities in Engineering
Review of the Literature and Overview of the Study

Engineers are different. Ingenious at devising and turning dreams into realities, they are the problem solvers of the scientific world (Tang, 2000). They are the drivers of progress, hewn from patterns of behavior grounded in modes of finding solutions (Camacho & Lord, 2013). They are 33% of Fortune 500 CEOs (Aquino, 2011), and 46% of 13 top CEOs surveyed (Williams, 2017), due to their attention to detail, analytical mindset, and training in systematic problem solving (Al-Saleh, 2014). While scientists are said to be driven by the need to know, engineers are driven by the need to solve problems; i.e., to do, to fix, to make work (Holtzapple & Reece, 2003). Scientists are concerned with fundamentals; engineers with design (Powell, 2006), and with an advantage born of greater math proficiency and learning by doing (Kokkelenberg & Sinha, 2010). This difference makes a case for studying engineering students in their own right. Yet, in most research studies, they are considered with other students in the sciences. Likewise, most research attention has been paid to minorities in STEM rather than minorities in engineering, which may blur important distinctions and policy implications (Atwaters, Leonard, & Pearson, 2015). This volume is devoted to minorities in engineering in anticipation of them being as different from minorities in STEM as engineers are from others in science.

Why focus on minorities in engineering? It could be said that engineering is made of up a conglomerate of minorities or that White students occupy a dwindling majority at the undergraduate level and a minority share at the masters and doctorate levels. Tang (2000) has argued that Asians should be treated as a minority, given treatment they face in the job market. The numbers of Middle Eastern and East Indian students who responded to the survey portion of this study beg for their own research attention. But we are concerned with the underrepresented minorities; i.e., African American, Hispanic, and Native American students. The concern is with increasing their numbers and facilitating the success of these students in engineering arise from several sources.

The traditional pool of the White male engineering workforce is decreasing due to low birth rates and interests in non-engineering careers such as business (Bonous-Hammarth, 2000; Brazziel & Brazziel, 1995, 2001). The general enrollments in science and engineering education are dwindling (Babco, 2001; Clewell, deCohen, Tsui, & Deterding, 2006; Gibbons, 2008; Ransom, 2015). The engineering workforce is becoming increasingly foreign-born, which has helped to fill the gap in educational enrollment and the shortfall in the American workforce (Chubin, May, & Babco, 2005). However, with more than 60% of the doctoral degrees in engineering going to non-Americans (Harvey & Anderson, 2005; Ryu, 2010), the wisdom of undue dependence on foreign talent has been questioned (Chubin, Donaldson, Olds, & Fleming, 2008; Chubin et al., 2005). For these reasons, there is an increased demand for the supply of home-grown engineering skill, and so the demand for minority participation has also increased. It then appears to be in the national *interest* to increase the representation of underrepresented minorities in engineering and to facilitate their success. This is not necessarily to say that it is a national *concern* to increase their numbers, or even a concern in academia. The effort to keep minorities out of engineering has had a long history (Johnson & Watson, 2005; Wharton, 1992), while Brazziel and Brazziel (1995) charged academic engineering with the most conservative resistance to inclusion they had encountered at the time. Slaton (2011) described the continued resistance to the inclusion of minorities as due in part to the belief that greater diversity means lower standards. Stewart, Malley, and Herzog (2016) cited a lack of concern with increasing female representation as reasons for their low numbers. Sax, Zimmerman, Blaney, Toven-Lindsey, and Lehman (2017) interviewed 15 department heads in Computer Science and found a concern with increasing the number of female students, but no efforts to increase the numbers of underrepresented minorities. Burack and Franks (2004) contend that it is less common for proponents of diversity to directly address the stubborn resistance to diversity that frequently prevails in the discipline, diversity being seen as threatening to the group identity; but they also contend that the special programs and places for minorities are not so much for the benefit of minorities as for the benefit of engineering. Indeed, with diversity also comes diversified perspectives, diversified creativities, and multiplied intelligences (Felder, Felder, & Dietz, 2002; Johnson & Watson, 2005)

The push to increase the numbers of underrepresented minorities has been supported by the vision of corporations and funding agencies (Pierre, 2015). As part of this effort, the National Action Council for Minorities in Engineering (NACME) has acted to marshal financial, educational, and support services to facilitate the participation of minorities in the engineering pipeline. Established in 1974, this 40-something year-old organization is guided by the goal to achieve parity in graduation rates (Pierre, 2015). The thrust has included the establishment of Minority Engineering Programs (MEPs) at scores of universities to develop the engineering pipeline (Landis, 2005). Along with that support has come an increased need to understand the factors that encourage minority involvement in engineering. Among its many activities, NACME provides scholarship support to 31 engineering schools with good records of retaining and graduating minorities. In addition to programmatic and scholarship support, NACME follows the progress of engineering students in its consortium schools (Lain & Frehill, 2012; Swail & Chubin, 2007). NACME's numerous reports have informed much of our knowledge of the participation levels of underrepresented African American, Latino and Native American students (NACME, 2010, 2014a, 2014b, 2014c).

Non-Minorities in Engineering

The concern for minority retention in engineering has called attention to the 50% dropout rate of non-minority students. Camacho and Lord (2013) have pointed out that the engineering dropout rate may be no higher than in other majors, but that the need for scientific talent and high qualifications of this group of students makes this a cause for concern. The dwindling numbers and high rate of failure have captured the attention of researchers searching for ways to lessen the departures. The rigors of the science curriculum present a formidable barrier to persistence. High failure rates in barrier courses such as calculus and physics have a strong negative impact on retention (Suresh, 2006/2007). Araque, Roldan, and Salguero (2009) argued that persistence in engineering requires a strong psychological profile for overcoming obstacles. Indeed, the personalities of engineering students have figured prominently in innovative teaching methods designed to increase passing rates (Felder et al., 2002).

What, then, is unique about the engineering personality? The inclination toward engineering appears to be a combination of factors, including a positive commitment to engineering, enjoyment of math and science, confidence in mathematical and scientific abilities, and an impersonal personality style (Besterfield-Sacre, Moreno, Shuman, & Atman, 2001; Capretz, 2003; Godfrey & Parker, 2010). Such characteristics fit the self-efficacy model proposed by Ambrose, Lazarus, and Nair (1998), who contend that the critical factor is a perceived ability to be efficacious in the field (e.g., Lee, Brozina, Amelink, & Jones, 2017). The psychology of the engineer is little discussed in the current retention literature, even though they are seen by themselves and others as different (NAE, 2008). Their primary mode of living is focused internally inside their own minds (Capretz, 2003). Their dominant approach to the world is as an organizer using methodical and pragmatic approaches to scientific research (Godfrey & Parker, 2010). As a group, they have adopted empirical problem solving and cognitive processing as a way of life (Scissons, 1979). They value intelligence, knowledge, and competence, and live in the world of ideas and strategic planning, giving them a tremendous value of and need for systems. They have difficulty expressing their internal images, since these are not readily translatable, and non-linear (Capretz, 2003). They may have little interest in other people's thoughts or feelings, are strongly field independent, organized, goal directed and with good study skills rather than being easy-going and exploratory, and are frequently misunderstood (Felder et al., 2002; Rosati, 2003; Scissons, 1984; Zimmerman et al., 2006). As is true for students in technical fields, engineering students were found to be less open to diversity than others in STEM as well as compared to all other majors (Ing & Denson, 2013).

Traditionally, it has been White male students who exemplified these characteristics and achieved a favored position in engineering. Further, such characteristics have been compatible with the traditional engineering curriculum and traditional methods of teaching (Felder et al., 2002). Thus, the majority of studies comparing minority and non-minority engineering students in college find that non-minorities fare better, with, as Camacho and Lord (2013) put it, the wind at their back. Indeed, majority students; i.e., males, have a higher expectancy of success in engineering (Lee et al., 2017). In the main, non-minorities exhibit better grades, with or without better test scores; better adjustment along many dimensions; and lower dropout rates (Araque et al., 2009; Borrego, Padilla, Zhang, Ohland, & Anderson, 2005; Chen & Weko, 2009; Suresh, 2006/2007). According to Marra, Rodgers, and Shen (2012), majority students less often cited academic reasons for leaving engineering compared to minority students, and majority students less often cited non-academic reasons such as feeling that they did not belong in engineering. Indeed, majority students do not typically share the problematic issues of belonging or comfort in the engineering milieu, since it is others (i.e., minorities and women) who must adapt to the White male fraternity (Baker, Tancred, & Whitesides, 2002; Besterfield-Sacre et al., 2001; Brown, Morning, & Watkins, 2005).

Much of what we know of engineering students comes from research describing engineering student learning styles. This work is concerned with broadening the traditional lecture teaching methods in engineering to include more active, hands-on methods that appeal to a wider range of students. These studies describe engineering students as comfortable with mathematics, introverted, intelligent, creative, effective, and good organizers or as single-minded, brusque, and not people-oriented (Broberg, Griggs, Paul & Lin, 2006; Godfrey & Parker, 2010; Scissons, 1979). Harrison, Tomblen, and Jackson (2006 [1955]) dispute the introverted stereotype and suggest they are simply impersonal. Nonetheless, introverted types are said to function well in the traditional lecture-teaching modes, but hands-on instruction—including project or problem-based instruction, hyper media, and web access—offer effective teaching modes for all learning styles (Zywno & Waalen, 2002). Virtually nothing is known about the personality orientation or learning styles of minority engineering students, although broadening teaching styles, such as group learning and problem- and project-based learning, have been shown to improve minority student performance as much or more than other students in the general sciences (Ro, Knight, & Loya, 2016).

Women in Engineering

Eighty-two percent of the undergraduate engineering degrees were awarded to men in 2009 (Camacho & Lord, 2013). Thus, the 18% of women who persisted to graduation occupy an extreme minority position. While other formerly male-dominated fields such as law and medicine have become nearly 50% female, engineering remains a bastion of maleness. While the intersectionality of gender, race, and ethnicity complicates the impact of gender on engineering expectations of efficacy in the field (Bruning, Bystydzienski, & Eisenhart, 2015), there is evidence that women in engineering achieve higher grade averages and have higher graduation rates than their male counterparts (Borrego et al., 2005; Brawner, Camacho, Lord, Long, & Ohland, 2012; Lord et al., 2009; Orr, Lord, Layton, & Ohland, 2014). Studies have suggested that the dearth of women in engineering is a problem of recruitment rather than retention, or as Lord et al. (2009) maintained, as a problem of low representation at matriculation. Su (2010) pointed out that the gender disparity is a problem specific to engineering, unlike other minorities who achieve low rates of overall degree attainment. Other work concluded that women were lacking in the math and science skills required for engineering (Blickenstaff, 2005). Adelman (1998) found that women took a heavier course load, and had difficulty coping with the fast pace and feelings of being overloaded by the engineering curriculum. His study found that women were less likely to graduate. Other work maintains that the small numbers stem from exclusionary attitudes and practices that begin early and negatively influence the degree of comfort that women can achieve in the engineering milieu. Fifolt and Abbott (2008), for example, cite the pervasiveness of the White male-dominated environment as a detriment to the recruitment of women (and minorities) in engineering programs. Indeed, the much-cited concept of "chilly climate," coined by Hall and Sandler (1984) and elaborated by Seymour and Hewitt (1997), has yielded a number of studies documenting the uncomfortable position of women in a conservative male bastion that can serve to reduce their commitment to engineering despite their often greater persistence (Clewell et al., 2006; Fouad, Singh, Fitzpatrick, & Lou, 2012; Gunter & Stambach, 2005; Wolf-Wendel, 2000). According to Blickenstaff (2005), women do lack preparedness for science, but also leave the field because of the male curricula and the chilly climate. Women experienced a more hostile or "incivil" environment (Miner, Diaz, & Rinn, 2017), endured a less comfortable existence (Margolis & Fisher, 2002; Seymour & Hewitt, 1997), expressed stronger needs for psycho-social support and mentoring (Fifolt & Abbott, 2008), reported different perceptions of fairness (Brainard, Metz, & Gilmore, 2000), received less encouragement (Baker et al., 2002), reported lower feelings of belonging (Marra et al., 2012; Meyer & Marx, 2014), and more disrespect (Camacho & Lord, 2013). According to work by Holland, Major, Morganson, and Orvis (2011), women have less access to informal capitalization activities that would promote their studies and careers, but gravitate to formal organizations frequented by other women. In other words, they were motivated by needs to avoid the sense of social isolation. Rosser (2012) found that female faculty members in the engineering academy (i.e., successful female engineers) reported histories of discomfort, alienation, and discrimination throughout their entire careers. Other literature has found that Black females and Latinas also showed a pattern of better performance and/or and higher graduation rates in engineering than their male counterparts (Ohland et al., 2011), despite lack of support (Leyva, 2016) and the discomfort of persistent micro-aggressions (Camacho & Lord, 2013). Relationships were important for women of color, both African American and Latino, with the points of critical influence being family support, university relationships, and industry experiences (DeCuir-Gunby, Grant, & Gregory, 2013; Leyva, 2016). As such, the effects of gender within race and ethnicity cannot be ignored in any study of minorities in engineering.

Minorities in Engineering

Enrollments are dwindling for engineering education in general, and they are dwindling for underrepresented minority students, as well (Malcolm-Piquex & Malcolm, 2015; Ransom, 2015). Nonetheless, underrepresented minority segments of the American population are the fastest growing; thus, they are potentially a resource for the development of scientific talent. Mathematics remains the premier pathway into engineering for minorities (Pearson & Miller, 2012), yet, these are also the groups that have fared least well academically, and their performance in mathematics as well as the sciences has been unencouraging (Dalton, Ingels, Downing, Bozick, & Owings, 2007; Grandy, 1997; Lutkus, Lauko, & Brockway, 2006). Further, dropout rates from engineering in general are not so much higher than from other majors, but too high given the need for scientific talent—around 50%; they are higher still for minorities and have remained so for decades. Morning and Fleming (1994) reported a 67% dropout rate, while their project was able to re-funnel minority dropouts into supportive engineering schools with a subsequent 67%

retention to graduation of these dropouts. In a study for NACME, Landis (2005) reported that 67% of minority freshmen in engineering drop out. In a report for seven cohorts at nine southeastern universities, including two Historically Black Colleges and Universities (HBCUs), Borrego et al. (2005) found dropout rates of 46%–73% for minorities. In a study for ASEE, Yoder (2012) reported dropout rates of 61.3% for African Americans, 55.6% for Hispanics, and 61.4% for Native Americans. NACME (2012) data indicated a dropout rate of 68.8% for African Americans. These dropout rates are all the more alarming because minority engineering students are among the most able students entering college.

Yoder (2012) suggested that primary reasons for the attrition of students from engineering include their perception of a learning environment that fails to motivate them and is unwelcoming; it is neither the students' capabilities nor their potential for performing well as engineers that determines their persistence. Marra et al. (2012) also blame more frequent departures due to feelings of not belonging in engineering, while Estrada, Woodcock, Hernandez, and Schultz (2011) reported that minority students have more difficulty integrating into the scientific community, despite strong self-efficacy in science. However, Meyer and Marx (2014) cited the inability to adapt to the rigors of engineering, as well as an impaired sense of belonging as reasons for leaving. Sondgeroth and Stough (1992) found that good as well as poor students complained of the hostile environment, although better students cited more strategies for coping with academics. Borrego et al. (2005) found lower GPA among minority leavers.

Ong, Wright, Espinosa, and Orfield (2011) maintain that there is a need for overarching theoretical frameworks in this area. According to Lewis (2003), a cogent theoretical understanding of minorities in science is hampered by too few studies, and this is more so the case for minorities in engineering because most studies lump engineering students with others in science. This common trend may blur important distinctions that may have implications for policy (Atwaters et al., 2015).

African Americans in Engineering

Recent data from NACME indicated that African Americans are still one of the most underrepresented groups in engineering (NACME, 2014a). While they are 12.3% of the population, they earn only 4.0% of the undergraduate degrees, represent 3.6% of the engineering workforce, and a mere 2.5% of the engineering faculty. Brown et al. (2005), Chubin et al. (2008), and Yoder (2012) found that African American engineering students were at much greater risk of dropping out than other students—due in part to the unwelcoming racial climate of engineering, and the perception of racism at Predominantly White Institutions (PWIs). The unwelcoming atmosphere has its roots in historical attempts to keep minorities out of engineering (Wharton, 1992), the role of the sciences in promulgating scientific racism (Pearson, 2005), and the continuing resistance to minority inclusion (Slaton, 2011).

Given the foregoing, retention to graduation is a high priority issue for African American students. Reichert and Absher (1997) stressed the importance of retaining students who have already expressed an interest in engineering. They noted that retention rates are not typically high among HBCUs, and that their place among the top producers of engineering graduates is due to their greater enrollment of African American students. Nonetheless, other studies have documented the fact that Black schools graduate a disproportionate share of engineering graduates (Chubin et al., 2005; Malcolm-Piquex & Malcolm, 2015; Ransom, 2015). These authors point out that top PWI producers of minority engineering graduates have dual degree programs with HBCUs, and thus possess a hidden Black college advantage (Malcolm-Piquex & Malcolm, 2015). Community colleges also offer a significant but unheralded pathway for underrepresented minorities into science and engineering. McPhail (2015) reported that 50% of Black and 55% of Hispanic science and engineering degree recipients in 2004 attended community colleges, while Adelman (1998) found that 20% of all engineering degree recipients began their careers at community colleges. Camacho and Lord (2013, p. 58) reported low dropout rates in engineering for transfer students from 11 institutions, many of whom came from community colleges—28%–36% for Latinos/as; 45%–51% for Black females and males, respectively. Students chose community colleges for reasons of finance, desire to be close to family, and smaller classes (Camacho & Lord, 2013; NACME, 2008). Strengthening of the community college transfer pipeline has been advocated (Plett, Lane, & Peter, 2016), along with strategies for improvement such as greater ease of articulation agreements (NACME, 2008) and enhanced contextualized project-based learning innovations (McPhail, 2015).

On demographics within engineering disciplines, Ohland, Lord, and Layton (2015) found that Black males and females were underrepresented in choosing and completing degrees in Civil Engineering. Unlike other groups, the dropouts were not replaced by transfers into the field. Black students were attracted to Electrical Engineering (while Mechanical Engineering attracted more Whites and Asians), and Black female transfer students were particularly successful in persisting to graduation in Electrical as well as Mechanical Engineering (Lord, Layton, Ohland, & Orr, 2013). Mechanical Engineering attracts the most students with higher graduation rates than other engineering disciplines, but Black students were the least likely to persist (Orr et al., 2014). Chemical Engineering did not show similar trends, but women graduated at rates equal to or better than those for males within ethnicity (Lord, Layton,

Ohland, Brawner, & Long, 2014). In this discipline, as well, Black students showed the lowest persistence rates.

Hispanic Students in Engineering

Hispanic students have shown the most progress in engineering participation, but they still remain underrepresented (NACME, 2014b). Their progress is attributable in part to their population growth. They are the fastest growing minority group with a population increase of 125% in the past 20 years (NCES, 2013; US Census Bureau, 2013), and with a potential college-age population (i.e., high school seniors) projected to increase by 40% in the near future (Prescott & Bransberger, 2012). Hispanic students constitute 17.0% of the population, but only 8.6% of the undergraduate degrees, 6.3% of the engineering workforce, and only 3.7% of engineering faculty. A majority of the advanced degrees earned come from two institutions: University of Puerto Rico and the University of Texas, El Paso (UTEP). While Hispanic females were almost non-existent in the pipeline in 1997, they constitute 40% of the graduates from University of Puerto Rico. However, the Latina percentage of undergraduate and master's degrees has remained stagnant for the last 10 years (NACME, 2011). Nevertheless, Latinas have exhibited equal or higher persistence rates than Latinos in several studies (Camacho & Lord, 2013; Lord et al., 2013, 2014; Ohland et al., 2011). Chubin et al. (2005) list the top producers of Hispanic students, which exist in states of high Hispanic concentrations, such as California and Texas.

Native Americans and Pacific Islanders

Native American and Alaska Natives have the smallest presence in the engineering pipeline (NACME, 2014c). They comprise 0.9% of the population, 0.4% of undergraduate degrees, 0.3% of the engineering workforce, and only 0.2% of engineering faculty. A large part of their small presence is due to a high school dropout rate that is twice that of White students (i.e., 117% higher). They, like other minorities, suffer from low degree attainment across major fields (Su, 2010). Also, unlike other minority women, Native American women exhibit persistence rates lower than their male counterparts (Lord et al., 2009). The top producers of Native American students occur in areas with high concentrations of Native people (Chubin et al., 2005). Unfortunately, this study will not shed much light on these students, because so few were study participants.

The Adjustment of Minorities in Engineering

In addition to the search for what might enhance the progress of minority students in engineering is the parallel question of how similar or different might minority engineers in college be compared to the general population of minority students in college.

The substantial literature on minority students in college strongly suggests that the issue of adjustment to the college environment is a critical one, and that race/ethnicity usually occasions less friendly treatment (Fries-Britt & Turner, 2002; Pascarella & Terenzini, 2005; Rovai, Gallien, & Wighting, 2005; Smith, 2009). Minority students exhibit poorer adjustment (Cabrera, Nora, Terenzini, Pascarella, & Hagedorn, 1999), and a weaker sense of belonging than non-minorities (Johnson et al., 2007), which in turn is said to account for their poorer performance and retention more so than differences in entering qualifications such as standardized test scores. The nature of their differential adjustment includes a more hostile experience (Karkouti, 2016), including the covert experience of micro-aggressions (Lewis, Mendenhall, Harwood, & Hunt, 2012; Solórzano, Ceja, & Yosso, 2000), greater psychological distress due to hostile incidents compared to non-minorities (Miner et al., 2017), less positive peer interactions, restricted extra-curricular involvement, and a less satisfactory social life (Smith & Moore, 2002). While the presence of covert and overt prejudice and racism colors the social interactions of minority students, experiences in the classroom and impoverished interactions with faculty are said to be more damaging largely because many non-minority professors harbor negative stereotypes about minority students' academic ability and potential that result in neglect (Ferguson, 2003; Jussim & Harber, 2005; Tennenbaum & Ruck, 2007).

Some studies show that Black and Hispanic students experience similar issues in predominantly White colleges, but differing experiences in Minority-Serving Institutions. Both groups are less academically successful; that is, have higher college attrition rates than non-minorities (Ishitani, 2006; Lee & Rawls, 2010), share similar college adjustment issues in the areas of academics, finances, personal/family relationships, and stress management (Constantine, Chen, & Ceesay, 1997; Constantine, Wilton, & Caldwell, 2003), experience adjustment issues requiring supportive relationships to maintain academic performance and personal well-being (Chiang, Hunter, & Yeh, 2004; Cole, 2008; Solberg & Viliarreal, 1997; Tomlinson-Clarke, 1998), and endure less positive teacher attitudes and expectations (Tennenbaum & Ruck, 2007). However, it has been suggested that Hispanic students have fewer ethnically related adjustment issues because they are less visible than African Americans (Ancis, Sedlacek, & Mohr, 2000; Oliver, Rodriguez, & Mickelson, 1985; Suarez-Balcazar, Orellana-Damacela, Portillo, Rowan, & Andrews-Guillen, 2003).

A long series of studies documents the better adjustment and outcomes for African American students in HBCUs (Allen, 1992; Constantine, 1995; Kim, 2002, 2004; Kim & Conrad, 2006; Shorette & Palmer, 2015). These include greater gains in cognitive skills (Fleming, 1984), more intellectual challenge, more faculty-student contact, more peer interaction (Palmer, Davis, & Maramba, 2010; Seifert, Drummond, & Pascarella, 2006), more positive quality of student-faculty relationships (Cokley, 2000; Palmer & Young, 2009), higher aspirations to doctoral level degrees (Freeman & McDonald, 2004), higher self-concept ratings (Berger & Milem, 2000), fewer racial identity conflicts (Fleming, 2002), and more support for high achievers (Fleming, 2004; Fries-Britt & Turner, 2002; Fries-Britt, 2004). Bridges, Kinzie, Laird, and Kuh (2008) cited a no-one-fails-here attitude, experimental pedagogy, service learning, and strategic freshmen seminars as factors in the Black college advantage, while Irvine and Fenwick (2011) maintained that Black teachers in HBCUs have higher expectations for Black students. Holland et al. (2011) found that a Black college was more supportive of capitalization activities promoting academic and career development than students in a predominantly White college, regardless of race of students, and that gender was less of an issue in an HBCU than in a PWI.

Hispanic students have shown some beneficial outcomes in Hispanic-Serving Institutions (HSIs), including stronger growth in academic self-concept (Cuellar, 2015) and larger gains in overall development (Laird, Bridges, Morelson-Quainoo, Williams, & Holmes, 2007). Other studies found no differences in outcomes such as social empowerment (Cuellar, 2015), perceptions of a supportive campus environment and faculty interactions (Laird et al., 2007), or beneficial outcomes to a lesser degree than found for Black students in Black schools (Bridges et al., 2008). Negative outcomes for students in HSIs have been reported, such as lower graduation rates (Contreras, Malcolm, & Bensimon, 2008), although controlling for critical student or institutional factors can eliminate graduation gaps (Flores & Park, 2015; Rodriguez & Galdeano, 2015). The attenuation in differential outcomes may also be due to the fact that HSIs are so designated by virtue of significant Hispanic enrollment, rather than a mission to uplift Hispanic students (Conrad & Gasman, 2015; Gasman, 2008; Nunez, Hurtado, & Galdeano, 2015). Furthermore, the outcomes for Hispanic students may depend on the exact percentage of Hispanic enrollment, a factor which shifts as does the number of HSIs in any given year (Rodriguez & Galdeano, 2015).

Studies of adjustment specific to minority students in engineering are still limited, but the adjustment factor may be no less critical for them and may well be more so because of the hostile climate and resulting ethnic isolation so often described in the literature (Brown et al., 2005; Landis, 2005; Miner et al., 2017). Studies do address the negative impact of the perception of racism on retention (Brown et al., 2005), more frequent departures due to feelings of not belonging in engineering (Marra et al., 2012), lower ratings of inclusiveness in the engineering environment (Lee, Matusovich, & Brown, 2014), reduced intellectual development (i.e., critical thinking) in White compared to Black engineering schools (Fleming, Garcia, & Morning, 1996), lack of support from instructors beginning in high school—the more gifted the student, the less perceived support (Fleming & Morning, 1998)—lower expectancies of success (Lee et al., 2017), and lower ratings of abilities compared to non-minorities in engineering (Ro & Loya, 2015). Meyerhoff Scholars who entered Ph.D. programs in engineering reported that instructors were as much a hindrance as help, that departments did not value diversity, that fellow White students were culturally insensitive, that fellow foreign students held hostile, negative stereotypes about Blacks, and that they were excluded from social groups (Maton, Watkins-Lewis, Beason, & Hrabowski, 2015). Comfort in the scientific milieu has been singled out as the most important factor in the adjustment of minorities in science, a factor twice as important as positive student-faculty contact (Person & Fleming, 2012). The greater benefits of attending HBCUs extend to females in STEM majors (Perna et al., 2009), as well as to students in engineering, whereas HBCUS are among the top producers of minority engineers (Chubin et al., 2005; Malcolm-Piquex & Malcolm, 2015; Ransom, 2015; Reichert & Absher, 1997). Certain PWI institutions that graduate large numbers of Black students have dual-degree relationships with HBCUs, thus providing a hidden Black college advantage (Malcolm-Piquex & Malcolm, 2015). Fleming, Smith, Williams, and Bliss (2013) found that for engineering students in two HBCUs, the benefits included challenging academic programs coupled with very caring faculty. Likewise, Hurtado et al. (2011) found greater faculty support for students in the sciences at HBCUs compared to more selective colleges. Thus, the usually hostile and competitive engineering environment was offset by strong interpersonal support.

Studies targeting Hispanic students in the science and engineering literature have found similar issues of adjustment challenges, difficult faculty interactions, and teamwork problems for them as well as both ethnic groups (Cole, 2008; Cole & Espinoza, 2008; Dika, 2012; Dika, Pando, & Tempest, 2016; Wolfe, Powell, Schlisserman, & Kirshon, 2016), including shared affirmative action stereotypes (Chou & Feagin, 2008). Millett and Nettles (2006) found no experiential differences between Hispanic and White doctoral students in STEM, despite lower publication and completion rates for Hispanic students. Minority students showed better academic outcomes with group learning structures (Ro et al., 2016). Positive interactions with faculty contributed to successful outcomes in engineering (Amelink & Meszaros, 2011; Vogt, 2008), but were especially critical for Black and Latino/a students, due to alleged social capital disadvantages (Cole & Espinoza, 2008; Hug & Jurow, 2013; Hurtado et al., 2011; Martin, Simmons, & Yu, 2013). Indeed, minority students' low expectancies of success can be combatted with positive interactions with faculty and staff (Martin et al., 2013). While routine interactions with faculty concerning grades may not lead to an academic advantage

(Cole & Espinoza, 2008), faculty mentorship and research with faculty members generally benefit minority students (Barlow & Villarejo, 2004; Kim & Sax, 2009; Maton & Hrabowski, 2004). Relationships with faculty and fellow peers were associated with retention among minority engineering students (Wao, Lee, & Borman, 2010). Black and Latino/a students may resist establishing contact with engineering faculty who are not of the same ethnicity, but interactions with peers and staff may provide a substitute for this form of support (Dika et al., 2016). Minorities and women in engineering appear to rely more on the social capital of other students rather than institutional support, perhaps because readier access to fellow students provides a needed sense of belonging (Wao et al., 2010). Fleming et al. (2013) found that engineering students at two HSIs reported strong support from faculty and peers that helped to cement their engineering identities. However, Flores, Navarro, Lee, and Luna (2014) found that Hispanic students in an HSI appeared to receive weaker reinforcement than White students in engineering, including the reinforcement provided by the strong representation of Whites in engineering. These authors suggested that Hispanic students in engineering have a less clear sense of self-efficacy in a non-traditional field to which they have less exposure. However, Wilson-Lopez, Mejia, Hasbún, and Kasun (2016) suggested that Latino/a students possess funds of knowledge compatible with engineering that effective curricula should draw upon; similar funds of knowledge exist for African American students (Barton & Tan, 2009).

In short, minorities in engineering appear to face similar conditions of unwelcome as minorities in other disciplines, but their adjustment issues have been studied in far less detail. Any similarities or differences between Black and Hispanic students appear to lack precise definition.

Programs for Minorities in Engineering

The establishment of Minority in Engineering Programs (MEPs) in scores of Engineering schools around the country was part of the coordinated effort of the corporate engineering community and funding agencies to increase the numbers of underrepresented minorities (e.g., Anderson-Rowland et al., 1999; Pierre, 2015). These multi-faceted programs were devoted to minority excellence in engineering and have enjoyed considerable success in strengthening skills and retaining students, largely through the development of supportive infrastructure (Landis, 2005; Swail & Chubin, 2007). While many of these programs have changed and evolved to serve all students (Shehab, Murphy, & Foor, 2012), the need for them has not lessened. NACME institutions are those with strong MEPs that foster the academic progress of minority students, in addition to other students, through staff support and multiple program components. Meeting the current challenges effectively requires an increasingly refined knowledge of factors that influence the development of minority talent, both at the individual level and the program level.

Swail and Chubin (2007) conducted the first analysis of the NACME block grant program that embraces schools with strong MEP programs that included ten institutions and noted significant variation among schools in academic outcomes. They called for quantification of program factors that might explain these variations. Acknowledging the resistance to minority inclusion, Slaton (2011) also calls for studies that show which sorts of programs, program components, and features of engineering education can yield greater minority participation, even in the current climate.

A number of successful programs in science, as well as engineering, have identified program ingredients critical in retaining and graduating minority students. These include a strong demand for excellence (AACU, 2007); strong faculty contact and/or mentoring (Walden & Foor, 2008); exposure to research experiences (Jones, Barlow, & Villarejo, 2010; Maton, Domingo, Stolle-McAllister, Zimmerman, & Hrabowski, 2009; Morning & Fleming, 2012); a strong structure for student support designed to break down barriers to inclusiveness (Anderson-Rowland, Urban, & Haag, 2000); a group-learning program structure that requires students to interact, work, and study together (Freeman, Alston, & Winborne, 2008; Garret-Ruffin & Martsolf, 2005; Ro et al., 2016); improving math competence (Hagedorn & DuBray, 2010; Simon & Farkas, 2008); summer bridge experiences (Morning & Fleming, 2012; Tsui, 2007); summer bridge experiences focused on problem solving to enhance mathematics competence (Fleming, 2012a); and scholarships (Fleming, 2012b; Hurtado et al., 2007; Kim & Otts, 2010).

Exposure to undergraduate research experience has now been established as a high-impact, highly effective strategy for encouraging college retention in general, as well as entrance into and retention in STEM fields in particular (Kuh, 2008; Lain & Frehill, 2012; Maton & Hrabowski, 2004; Maton et al., 2009). Several studies have specifically demonstrated that research experiences are beneficial for minority students in general and in the sciences (Morning & Fleming, 2012; Foertsch, Alexander, & Penberthy, 2000; Hurtado et al., 2008; Jones, Barlow, & Villarejo, 2010; Kuh, 2008). The Louis Stokes Alliance for Minority Participation in Science, Technology, Engineering, and Mathematics (LSAMP) program established at universities nationwide has involved minority students in undergraduate research with promising outcomes (Fleming, 2017; Morning & Fleming, 2012; Pearson & Marshall, 2016). Lain and Frehill (2012) found in a study of NACME graduating scholars that research experiences were a motivating factor in their retention to graduation.

While many programmatic efforts deserve detailed attention, three are highlighted here. The Pre-Engineering programs coordinated by the University of Akron were modeled after the college-level MEP programs described in the works of Raymond Landis (2005). Beginning in 1993, the College of Engineering aimed to enhance the possibility of academic success in STEM fields among under-represented high school students. Their concept of success relied on five general guidelines (Lam, Doverspike, & Mawasha, 1997): (1) math and science knowledge; (2) commitment to a STEM career; (3) educational and occupational values and beliefs; 4) social support, including role models, peer, faculty, and family support; and (5) self-efficacy or competency. An overarching guideline was that selected students must have demonstrated some interest in or potential for math and science. The heart of the effort is a six-week residential summer program built around restructured pedagogy from lecture to inquiry-based approach, including a series of hands-on design projects where students could engage in real-world engineering activities without the fear of having the wrong answers. The summer program was supported by academic year activities such as: (1) diagnostic testing; (2) orientation; (3) peer mentoring; (4) study skill and group dynamic workshops; (5) bridge-up classes; and (6) academic advising. Periodic evaluations of the College of Engineering programs were conducted (e.g., Lam, Doverspike, & Mawasha, 1997, 1999; Lam, Mawasha, & Chu, 1994; Lam, Ugweje, Mawasha, & Srivatsan, 2003), and a ten-year evaluation by Lam, Srivatsan, Doverspike, Vesalo, and Mawasha (2005) documented significant increases in GPA: a 98% college entry and a 66% STEM entry rate. Other successful pre-college engineering programs in the Ohio area using a similar program model have been reported by Gilmer (2007) and Yelamarthi and Mawasha (2008).

Project Preserve was a program uniquely designed to incorporate program components that were true to research evidence (Fleming, 2012a; Fleming et al., 1996; Fleming & Morning, 1998; Morning & Fleming, 1994). After a thorough review of the minorities in engineering retention literature, three critical program components were designed to ensure: (1) bonding to the university through living in on-campus housing, campus-based jobs, and participation in organized MEP programs and student professional organizations; (2) interpersonal bonding through required weekly meetings with a faculty member individually and in groups; and (3) student needs for intellectual development fulfilled by an emphasis on cognitive growth through explicit strategic activities that included "Stress on Analytical Reasoning," "Learning to Learn," "Efficacy," and "Problem Solving an Comprehension." The project targeted 100 students who had dropped out of engineering and rechanneled to three institutions known to provide strong supportive programs for minority engineering students: California State University at Northridge; The City College of New York; and Xavier University of New Orleans, an HBCU. A comprehensive evaluation documented progress in achieving the three program goals, and the successful retention to graduation of 67% of those students who would have left the engineering pipeline.

There were variations in the results such that results were best at the HBCU and worst where on-campus living or faculty involvement could not be achieved.

The Meyerhoff Scholars Program is perhaps the best known and most successful program for minorities in undergraduate STEM education. Housed at the University of Maryland, Baltimore County, it has been in existence since 1988. While at first only open to African American students, its mission has widened to students of all backgrounds. The program is comprehensive in its interventions, but also engages in repeated evaluations to refine its operations (Baker, 2006; Maton & Hrabowski, 2004; Maton, Hrabowski, & Schmitt, 2000; Maton et al., 2009, 2015). In addition to its overall guiding set of four critical factors—(1) knowledge and skills, (2) motivation and support, (3) monitoring and advising, and (4) academic and social integration—are specific program components of financial aid, summer bridge, study groups, program values, program community, staff academic advising and personal counseling, tutoring, summer research internships, faculty involvement, peer mentoring, community service, and administrative involvement. African American Meyerhoff students were 2.5 times more likely to enter Ph.D. programs in engineering compared to a control group. Highly rated program components by students were demand for achievement striving, scholarships, community belonging, summer bridge, and study groups. GPA and number of research internships were the best predictors of postcollege entry into graduate engineering programs.

While the general landscape of what constitutes effective programs has been mapped, a delineation of program features that specifically serve minorities in engineering is still lacking.

From the foregoing review, it appears that:

1. Despite the excellent qualifications of engineering students, their dropout rates are high, hovering around 50% (which is much like the general population of college students), but the rates are higher still for minority students hovering around 67%. Among minority students, these rates vary according to unspecified factors except that rates are lower in HBCUs.
2. For minority students, the low representation in engineering seems due to low numbers at entry, as well as high dropout rates. In contrast, for women, low representation appears due to low numbers at entry since women typically persist in engineering as well or better than males. There is some evidence that this pattern is also true for minority women.
3. It appears that minorities in engineering face some of the same adjustment problems as do minorities in the general college adjustment literature. However, these issues for minorities in engineering have been studied in far less detail.

4. Effective program support has been a hallmark of the Minorities in Engineering effort, and the ingredients of successful programs in the sciences have been widely reported. However, there is still a need for a precise delineation of *factors*, in order of importance, for minorities in engineering.
5. The issues for minorities may or not be similar to those for women, and the issues for the intersection of women and ethnicity need further clarification in engineering.

Overview of the Study

The National Action Council for Minorities in Engineering, an umbrella organization for the distribution of scholarships to disadvantaged minority students in undergraduate engineering education, has spearheaded a three-pronged study designed to identify factors enhancing the academic success of these students.

The first step was to conduct institutional analyses of academic performance and retention to graduation data in NACME member institutions to assess the academic context in which the target students operated, as well as to compare the findings with other reports. The question for this study was: are these rates the same or different in NACME member institutions, chosen for their stronger programming and support for minorities in engineering? This study of institutional success factors for minorities was designed not to be a study of the best or most prestigious engineering schools with the best students as judged by entering standardized test scores, but of effective schools that do the most with what they have. The concern was with what factors appear to distinguish some of the most effective engineering schools.

The second step was to get to know the minority engineering students well enough to construct profiles of them and to design a study of their engineering education experiences. To do this, focus groups were conducted, and mini-surveys were administered during the interviews for additional analysis.

The third step was to conduct an online survey designed to assess the program participation and adjustment factors that facilitate the success of minority engineers in college. These factors were examined as a function of GPA, of matriculation in more effective schools, and classification, that is the value-added of surviving to the senior years.

Chapter 1 describes the history of the National Action Council for Minorities in Engineering, its refocused purposes, and its place in the Minorities in Engineering effort as an umbrella for the distribution of scholarship funds to member institutions, and as a monitoring agency for the progress of underrepresented minorities in the engineering landscape.

Chapter 2 provides a capsule review of the literature relevant to the study's purposes. It begins with the premise that engineers are different from other scientists in definable ways and raises the issue of whether minority engineering students are also different from others, and particularly from other minorities in college. It explores statistics on underrepresented minorities of African American, Hispanic, and Native American students, although Native Americans will receive limited treatment in this volume. A summary of the status of non-minorities in engineering is followed by the adjustment issues faced by women in engineering, and by minorities in engineering in the larger context of minorities in college. The review ends with a description of the salient features of effective programs for minorities in engineering, followed by the issues that comprise the need for this study.

Chapter 3 tackles an aggregate comparative institutional analysis of minority and non-minority academic performance and retention to graduation at 28 of the NACME Block Grant institutions. Relevant data was requested and provided by each of the schools for categories of each group of students. After test scores were controlled, there were no differences between minority and non-minority groups in 1-year retention and 4-year or 6-year graduation rates. This suggests that other studies have compared students of unequal abilities, and that comparing students of roughly equal abilities may add another dimension to the analysis of engineering education progress. Comparisons of academic performance, however, did reveal evidence of minority underperformance, particularly in basic science courses. Minority women achieved better grades than their male counterparts, but this academic advantage did not pay off in higher graduation rates for them, as it typically does for women, minority or otherwise, in previous studies.

Chapter 4 reports the findings from focus groups of undergraduate engineering students in 11 institutions. They were asked to discuss how they gravitated to engineering; their relationship to mathematics; their experiences with professors, with prejudice and racism; and their career goals and plans for the future. In all of the conversations, the most notable omissions were complaints. Compared to most minority students that we have studied, and conversed with in similar focus groups, these students did not complain about the teaching, the professors, racism, or anything else. They treated all of their issues—academic, social, and racial—as problems to be solved, for that is what they do.

Chapter 5 presents an analysis of the mini-survey administered during the focus group sessions. The survey asked four major questions and presents the most frequently given responses. These responses were then examined as a function of student success measures; that is, by reported GPA, by reported test scores, by longevity or classification, and by university caliber, measured by average test scores for each institution. The results suggested that major success factors for these students were exposure to the hands-on work of engineering including problem- or project-based courses and internships, and a pro-active attitude toward school, work, and their futures.

Chapter 6 describes the results of online surveys on the undergraduate engineering experience that were completed by 632 minority students (largely Black and Hispanic) from 18 of the NACME institutions, of whom 40% were female, to determine the role that program participation, college adjustment, and background play in minority engineering student success. A factor analytic sorting was carried out using all independent measures significantly associated with three dependent measures of "success:" overall GPA, matriculation in schools more effective in graduating minority students, and retention to the senior year. The results suggested that a combination of faculty/college satisfaction, academic management skills, and Minority Engineering Program participation contributed most to minority success in engineering.

The study also included 513 non-minority students, 35% of whom were female. The major difference between minority and non-minority success factors was that program participation in general was a negative factor for non-minorities, engaged in by less successful students; the exceptions to this rule were participation in internships and research. Non-minority students had significantly higher test scores, but grades that were only slightly better than minorities, given similar ability. Minority students reported better adjustment than non-minorities on a range of measures that appears due to immersion in MEP programming.

Chapter 7 describes the survey results for 260 Black and 368 Hispanic students in 18 NACME Institutions. The results showed that for both groups of students, a combination of academic skill sets, satisfaction with the engineering environment, and participation in critical program components, such as internships, contributed most to three measures of "success" in a series of statistical analyses. Black students participated more in MEP program components, and their success depended more on program involvement. Matriculation in an HBCU contributed significantly to enhanced 6-year graduation rates for Black students. Compared to Black students, Hispanic students reported a stronger scientific orientation, while their success relied less on MEP program involvement and more on managerial skill, and interpersonal interactions of several kinds—with peers, in study groups, and with faculty. Attending an HSI had no effect on engineering success for these students. Gender differences among Black students were minor, while among Hispanic students, female students participated more in MEP program offerings, and were more outgoing.

Chapter 8 engages in an exhaustive dissection of gender and ethnicity because previous survey analyses showed only minor gender differences. Citing the many disadvantages and adjustment problems of women in engineering in previous studies, the search was on for similar issues here, with a resulting examination of gender differences in each of the five ethnic groups represented. However, none were found. Although women in the focus groups did report sexist treatment, the female survey respondents showed no evidence that their performance or adjustment was undermined by being female. To the contrary, women performed as well or better than men, and Asian women outperformed all others. Women generally reported better adjustment which may be linked to their participating more in academic program offerings. Their better social adjustment may indicate better social skills, and/or a greater reliance on the social capital of others.

Chapter 9 provides an examination of Black students in Black and White engineering institutions. A long history of studies has found huge advantages for Blacks in Black schools, and this study sought to examine the survey evidence for similar differences. Previous chapters found HBCUs among the most effective institutions in graduating minority students and established HBCU matriculation as an independent success factor. Focused comparisons found that Black students scored higher on a number of adjustment measures in Black schools, and that the differences were more pronounced the longer they were in these schools. On the other hand, Blacks in White engineering schools reported more opportunity advantages, such as greater financial support, greater likelihood of being NACME Scholars, and greater access to undergraduate research opportunities. However, there were no differences in access to the all-important internship participation. Despite differences in institutional experience, program participation was the key to success for all students, regardless of institutional type.

Chapter 11, a postscript on NACME Scholars who were able to identify themselves showed that they distinguish themselves in important ways—by higher academic performance and greater participation in the critical MEP programs of research and internships. Black female Scholars were the most likely to participate in internships, while Black Scholars in general reported the greatest comfort in the scientific milieu.

Chapter 10 summarizes the study findings and comes to five major conclusions. First, in engineering, ability is paramount. In this case, ability was measured by standardized test scores. No matter how one stands on testing, these scores were important to student success in most cases. That minorities have lower test scores in the institutional study, as well as the student survey, is a problem that should be addressed.

Second, the ability to manage one's ability and the academic environment seems to take a close second to ability in achieving success. Such managerial abilities include the meta-analytic organization of information, the protection of concentration, and the assessment of faculty, as well as effort and time management.

Third, MEP program participation occupies a central position in minority student success. It appears to enable good minority student adjustment, better than that observed for non-minority students. The most critical programs were Internships, Research, Study Groups, Middle School STEM Programs, Peer Mentoring, Tutoring, and Freshman Orientation Courses.

Fourth, attending an HBCU constitutes a pathway to engineering success in its own right. There was some evidence that MEP programming in Predominantly White Institutions offers an alternative to what an HBCU provides for its students; i.e., better academic adjustment. This conclusion is consistent with a great deal of prior research, as well as research on students in STEM disciplines that the HBCU difference is better academic integration rather than social integration, and strong student programming informed by retention theory. Regrettably, there was no evidence that HSIs served this function for Hispanic students in this study.

Fifth, again and again, the study provides evidence that hands-on exposure to the work of engineering—as in problem- or project-based courses, research participation, and particularly industry internships—constitute primary success factors.

The final Chapter 12 places the study's findings in the context of program operations and needed actions at its member institutions. Primary among the implications are the need to strengthen the hands-on aspects of support programs; in particular, increasing the availability of internship placements and undergraduate research experiences for minority students were clearly indicated.

References

AACU (Association of American Colleges and Universities). (2007). *College learning for the new global century: A report from the national leadership council for liberal education and America's promise*. Washington, DC: Author.

Adelman, C. (1998). *Women and men of the engineering path: A model for analyses of undergraduate careers*. Washington, DC: U.S. Department of Education. US Government Printing Office, Superintendent of Documents, Mail Stop: SSOP, Washington, DC 20402-9328. ED 419 696.

Allen, W. R. (1992). The color of success: African American college student outcomes at predominantly White and historically Black public colleges and universities. *Harvard Educational Review*, 62(1), 26–44.

Al-Saleh, Y. (2014). Why engineers make great CEOs. *Forbes*, May 29.

Ambrose, S., Lazarus, B., & Nair, I. (1998). No universal constants: Journeys of women in engineering and computer science. *Journal of Engineering Education*, 87(4), 363–368.

Amelink, M., & Meszaros, P. (2011). A comparison of educational factors promoting or discouraging the intent to remain in engineering by gender. *European Journal of Engineering Education*, 36(1), 47–62.

Ancis, J. R., Sedlacek, W. E., & Mohr, J. J. (2000). Student perceptions of campus climate by race. *Journal of Counseling and Development*, 78(2), 180–186.

Anderson-Rowland, M., Blaisdell, S. L., Fletcher, S. L., Fussell, P. A., McCartney, M. A., Reyes, M. A., & White, M. A. (1999). A collaborative effort to recruit and retain underrepresented engineering students. *Journal of Women and Minorities in Science and Engineering, 5*(4), 323–349.

Anderson-Rowland, M. R., Urban, J. E., & Haag, S. G. (2000). Including engineering students. In *Frontiers in Education Conference, 2000*. Kansas City, MO. *FIE 2000. 30th Annual* (Vol. 2, pp. F2F-5). New York: IEEE.

Aquino, J. (2011). 33% of CEOs majored in engineering—and the surprising facts about your boss. *Business Insider*, March 23. Retrieved from www.businessinsider.com/ceos-majored-in-engineering-2011-3.

Araque, F., Roldan, C., & Salguero, A. (2009). Factors influencing university dropout rates. *Computers and Education, 53*(3), 563–574.

Atwaters, S. Y., Leonard, J. D., & Pearson, W. (2015). Beyond the Black-White minority experience: Undergraduate engineering trends among African Americans. In J. B. Slaughter, Y. Tao, & W. Pearson, Jr. (Eds.), *Changing the face of engineering: The African American experience* (pp. 149–188). Baltimore, MD: Johns Hopkins University Press.

Babco, E. L. (2001). *Underrepresented minorities in engineering: A progress report*. Washington DC: American Association for the Advance of Science & Alliances for Graduate Education and the Professoriate.

Baker, E. (2006). Meyerhoff scholarship program. In M. J. Cuyjet (Ed.), *African American men in college*. San Francisco: Jossey-Bass.

Baker, S., Tancred, P., & Whitesides, S. (2002). Gender and graduate school: Engineering students confront life after the B.Eng. *Journal of Engineering Education, 91*(1), 41–48.

Barlow, A. E. L., & Villarejo, M. (2004). Making a difference for minorities: Evaluation of an educational enrichment program. *Journal of Research in Science Teaching, 41*(9), 861–881.

Barton, A. C., & Tan, E. (2009). Funds of knowledge and discourses and hybrid space. *Journal of Research in Science Teaching, 46*(1), 50–73.

Berger, J. B., & Milem, J. F. (2000). Exploring the impact of historically Black colleges in promoting the development of undergraduates' self-concept. *Journal of College Student Development, 41*(4), 381–394.

Besterfield-Sacre, M., Moreno, M., Shuman, L. J., & Atman, C. J. (2001). Gender and ethnicity differences in freshmen engineering student attitudes: A cross-institutional study. *Journal of Engineering Education, 90*(4), 477–490.

Blickenstaff, J. C. (2005). Women and science careers: Leaky pipeline or gender filter? *Gender and Education, 17*(4), 369–386.

Bonous-Hammarth, M. (2000). Pathways to success: Affirming opportunities for science, mathematics and engineering majors. *Journal of Negro Education, 69*(1/2), 92–111.

Borrego, M. J., Padilla, M. A., Zhang, G., Ohland, M. W., & Anderson, T. J. (2005). *Graduation rates, grade-point average, and changes of major of female and minority students entering engineering*. Proceedings, ASEE/IEEE Frontiers in Education Conference, October 19–22, Indianapolis, IN.

Brainard, S. G., Metz, S. S., & Gilmore, G. M. (2000). WEPAN pilot climate survey: Exploring the environment for undergraduate engineering students. Women in Engineering Programs and Advocates Network. In *Technology and Society, 1999. Women and Technology: Historical, Societal, and Professional Perspectives* (pp. 61–72). New York NY: IEEE. Retrieved from www.wepan.org.

Brawner, C. E., Camacho, M. M., Lord, S. M., Long, R. A., & Ohland, M. W. (2012). Women in industrial engineering: Stereotypes, persistence, and perspectives. *Journal of Engineering Education, 101*(2), 288–318.

Brazziel, W. F., & Brazziel, M. E. (1995). Broadening the search for minority science and engineering doctoral starts. *Journal of Science Education and Technology, 4*(2), 141–149.

Brazziel, W. F., & Brazziel, M. E. (2001). Factors in decisions of underrepresented minorities to forgo science and engineering doctoral study: A pilot study. *Journal of Science Education and Technology, 10*(3), 273–281. doi: 10.1023/A:1016694701704.

Bridges, B. K., Kinzie, J., Laird, T. F. N., & Kuh, G. D. (2008). Student engagement and student success at historically Black and Hispanic-serving institutions. In M. Gasman, B. Baez, & C. S. V. Turner (Eds.), *Understanding minority-serving institutions* (pp. 217–236). Albany, NY: State University of New York Press.

Broberg, H., Griggs, K., Paul, I., & Lin, H. (2006). Learning styles of electrical and computer engineering technology students. *Journal of Engineering Technology, 23*(1), 40.

Brown, A. R., Morning, C., & Watkins, C. (2005). Influence of African American engineering student perceptions of campus climate on graduation rates. *Journal of Engineering Education, 94*(4), 263–271.

Bruning, M. J., Bystydzienski, J., & Eisenhart, M. (2015). Intersectionality as a framework for understanding diverse young women's commitment to engineering. *Journal of Women and Minorities in Science and Engineering, 21*(1), 1–26.

Burack, C., & Franks, S. (2004). Telling stories about engineering: Group dynamics and resistance to diversity. *NWSA Journal, 16*(1), 79–95.

Cabrera, A. F., Nora, A., Terenzini, P. T., Pascarella, E. T., & Hagedorn, L. S. (1999). Campus racial climate and the adjustment of students to college: A comparison between white students and African American students. *Journal of Higher Education, 70*(2), 134–160.

Camacho, M. M., & Lord, S. M. (2013). *The borderlands of education: Latinas in engineering*. Lanhan, MD: Lexington Books.

Capretz, L. F. (2003). Personality types in software engineering. *International Journal of Human-Computer Studies, 58*(2), 207–214.

Chen, X., & Weko, T. (2009). *Students who study science, technology, engineering and math (STEM) in postsecondary education*. Washington, DC: U.S. Department of Education.

Chiang, L., Hunter, C. D., & Yeh, C. J. (2004). Coping attitudes, sources, and practices among Black and Latino college students. *Adolescence, 39*(156), 793–815.

Chou, R., & Feagin, J. (2008). *The myth of the model minority: Asian Americans facing racism*. Boulder, CO: Paradigm Publishers.

Chubin, D. E., Donaldson, K., Olds, B., & Fleming, L. (2008). Educating generation net: Can U.S. engineering woo and win the competition for talent? *Journal of Engineering Education, 97*(3), 245–258.

Chubin, D. E., May, G. S., & Babco, E. L. (2005). Diversifying the engineering workforce. *Journal of Engineering Education, 94*(1), 73–86.

Clewell, B. C., deCohen, C. C., Tsui, L., & Deterding, N. (2006). *Revitalizing the nation's talent pool in STEM: Science, technology, engineering and math.* Washington, DC: Urban Institute.

Cokley, K. (2000). An investigation of academic self-concept and its relationship to academic achievement in African American college students. *Journal of Black Psychology, 26*(2), 148–164.

Cole, D. (2008). Constructive criticism: The role of student-faculty interactions on African American and Hispanic students' educational gains. *Journal of College Student Development, 49*(6), 587–605. doi: 10.1353/csd.0.0040.

Cole, D., & Espinoza, A. (2008). Examining the academic success of Latino students in science technology engineering and mathematics (STEM) majors. *Journal of College Student Development, 49*(4), 285–300. doi: 10.1353/csd.0.0018.

Conrad, C., & Gasman, M. (2015). *Educating a diverse nation: Lessons from minority-serving institutions.* Cambridge, MA: Harvard University Press.

Constantine, J. M. (1995). The effect of attending historically Black colleges and universities on future wages of Black students. *Industrial & Labor Relations Review, 48*(3), 531–546.

Constantine, M. G., Chen, E. C., & Ceesay, P. (1997). Intake concerns of racial and ethnic minority students at a university counseling center: Implications for developmental programming and outreach. *Journal of Multicultural Counseling and Development, 25*(3), 210–218.

Constantine, M. G., Wilton, L., & Caldwell, L. D. (2003). The role of social support in moderating the relationship between psychological distress and willingness to seek psychological help among Black and Latino college students. *Journal of College Counseling, 6*(2), 155–165.

Contreras, F. E., Malcolm, L. E., & Bensimon, E. M. (2008). Hispanic-serving institutions: Closeted identity and the production of equitable outcomes for Latina/o students. In M. Gasman, B. Baez, & C. S. V. Turner (Eds.), *Understanding minority-serving institutions* (pp. 71–90). Albany, NY: State University of New York Press.

Cuellar, M. (2015). Latina/o student characteristics and outcomes at four-year Hispanic-serving institutions (HSIs), emerging HSIs, and Non-SHIs. In A. M. Nunez, S. Hurtado, & E. C. Galdeano (Eds.), *Hispanic-serving institutions: Advancing research and transformative practice* (pp. 101–120). New York: Routledge.

Dalton, B., Ingels, S. J., Downing, J., Bozick, R., & Owings, J. (2007). *Advanced mathematics and science coursetaking in the spring high school senior classes of 1982, 1992 and 2004: Statistical analysis report.* Washington, DC: U.S. Department of Education.

DeCuir-Gunby, J. T., Grant, C., & Gregory, B. B. (2013). Exploring career trajectories for women of color in engineering: The experiences of African American and Latina engineering professors. *Journal of Women and Minorities in Science and Engineering, 19*(3), 209–225.

Dika, S. L. (2012). Relations with faculty as social capital for college students: Evidence from Puerto Rico. *Journal of College Student Development, 53*(4), 596–610.

Dika, S. L., Pando, M. A., & Tempest, B. (2016). *Investigating the role of interaction, attitudes, and intentions for enrollment and persistence in engineering among underrepresented minority students*. Proceedings of the 2016 Annual Conference of the American Society of Engineering Education (ASEE), New Orleans, Louisiana. Retrieved from https://peer.asee.org/17069.

Estrada, M., Woodcock, A., Hernandez, P. R., & Schultz, P. (2011). Toward a model of social influence that explains minority student integration into the scientific community. *Journal of Educational Psychology*, *103*(1), 206.

Felder, R. M., Felder, G. N., & Dietz, E. J. (2002). The effects of personality type on engineering student performance and attitudes. *Journal of Engineering Education*, *91*(1), 3–17.

Ferguson, R. F. (2003). Teachers' perceptions and expectations and the Black-White test score gap. *Urban Education*, *38*(4), 460–507.

Fifolt, M. M., & Abbott, G. (2008). Differential experiences of women and minority engineering students in a cooperative education program. *Journal of Women and Minorities in Science and Engineering*, *14*(3), 253–267.

Fleming, J. (1984). *Blacks in college*. San Francisco: Jossey-Bass.

Fleming, J. (2002). Identity and achievement: Black ideology and the SAT in African American college students. In W. R. Allen, M. B. Spencer, & C. O'Connor (Eds.), *African American education: Race community, inequality and achievement* (pp. 77–92). Burlington, MA: Elsevier Science.

Fleming, J. (2004). The significance of historically Black colleges for high achievers: Correlates of standardized test scores in African American students. In M. C. Brown II & K. Freeman (Eds.), *Black colleges: New perspectives on policy and practice* (pp. 29–52). Westport, CT: Praeger.

Fleming, J. (2012a). Problem solving, critical thinking, and academic performance: Summer bridge for entering minority engineering students at California State University at Northridge. In J. Fleming, *Enhancing minority student retention and academic performance: What we can learn from program evaluations* (pp. 62–73). San Francisco: Jossey-Bass.

Fleming, J. (2012b). Enhancing learning in science programs at the City College of New York. In J. Fleming, *Enhancing minority student retention and academic performance: What we can learn from program evaluations* (pp. 86–101). San Francisco: Jossey-Bass.

Fleming, J. (2017). *City College of the City University of New York Louis Stokes alliance for minority participation: Evaluation report*. Unpublished manuscript. New York: City College New York.

Fleming, J., Garcia, N., & Morning, C. (1996). The critical thinking skills of minority engineering students: An exploratory study. *Journal of Negro Education*, *64*(4), 437–453.

Fleming, J., & Morning, C. (1998). Correlates of the SAT in minority engineering students: An exploratory study. *Journal of Higher Education*, *69*(1), 89–108.

Fleming, L. N., Smith, K. C., Williams, D. G., & Bliss, L. B. (2013). *Engineering identity of Black and Hispanic undergraduates: The impact of minority serving institutions*. Proceedings of American Society for Engineering Education Annual Conference and Exposition (ASEE), Atlanta, GA, pp. 23.510.1–23.510.18.

Flores, L. Y., Navarro, R. L., Lee, H. S., & Luna, L. L. (2014). Predictors of engineering-related self-efficacy and outcome expectations across gender and racial/ethnic groups. *Journal of Women and Minorities in Science and Engineering*, *20*(2), 149–169.

Flores, S. M., & Park, T. J. (2015). The effect of enrolling in a minority-serving institution for Black and Hispanic students in Texas. *Research in Higher Education, 56*(3), 247–276.

Foertsche, J., Alexander, B. B., & Penberthy, D. (2000). Summer research opportunity programs (SROPs) for minority undergraduates: A longitudinal study of program outcomes, 1986–1996. *Council of Undergraduate Research Quarterly, 20*(3), 114–119.

Fouad, N. A., Singh, R., Fitzpatrick, M. E., & Lou, J. P. (2012). *Stemming the tide: Why women leave engineering.* Milwaukee, WI: University of Wisconsin Milwaukee Center for the Study of the Workplace. Retrieved from http://studyofwork.com/files/2012/10/NSF_Report_2012-101d98c.pdf.

Freeman, K. E., Alston, S. T., & Winborne, D. G. (2008). Do learning communities enhance the quality of students' learning and motivation in STEM? *Journal of Negro Education, 77*(3), 227–240.

Freeman, K., & McDonald, N. (2004). Attracting the best and brightest: College choice influences at Black colleges. In M. C. Brown II & K. Freeman (Eds.), *Black colleges: New perspectives on policy and practice* (pp. 53–64). Westport, CT: Praeger.

Fries-Britt, S. (2004). The challenges and needs of high-achieving Black college students. In M. C. Brown II & K. Freeman (Eds.), *Black colleges: New perspectives on policy and practice* (pp. 161–175). Westport, CT: Praeger.

Fries-Britt, S., & Turner, B. (2002). Uneven stories: The experiences of successful Black collegians at a historically Black and a traditionally White campus. *Review of Higher Education, 25*(3), 315–330.

Garret-Ruffin, S., & Martsolf, D. S. (2005). Development, implementation, and evaluation of a science learning community for underrepresented students. *Journal of Women and Minorities in Science and Engineering, 11*(2), 197–208.

Gasman, M. (2008). Minority-serving institutions: A historical backdrop. In M. Gasman, B. Baez, & C. S. V. Turner (Eds.), *Understanding minority-serving institutions* (pp. 18–27). Albany, NY: State University of New York Press.

Gibbons, M. T. (2008). *Engineering by the numbers: Engineering degrees in the U.S.* New York: American Society for Engineering Education.

Gilmer, T. C. (2007). An understanding of the improved grades, retention and graduation rates of STEM majors at the Academic Investment in Math and Science (AIMS) Program of Bowling Green State University (BGSU). *Journal of STEM Education: Innovations and Research, 8*(1/2), 11–21.

Godfrey, E., & Parker, L. (2010). Mapping the cultural landscape in engineering education. *Journal of Engineering Education, 99*(1), 5–22.

Grandy, J. (1997). Gender and ethnic differences in the experiences, achievements, and expectations of science and engineering majors. *Journal of Women and Minorities in Science and Engineering, 3*(3), 199–143.

Gunter, R., & Stambach, A. (2005). Differences in men and women scientists' perceptions of workplace climate. *Journal of Women and Minorities in Science and Engineering, 11*(1), 97–116.

Hagedorn, L. S., & DuBray, D. (2010). Math and science success and nonsuccess: Journeys within the community college. *Journal of Women and Minorities in Science and Engineering, 16*(1), 31–50.

Hall, R. M., & Sandler, B. R. (1984). *Out of the classroom: A chilly campus climate for women?* Washington, DC: Association of American Colleges.

Harrison, R., Tomblen, D. I. & Jackson, T. A. (2006 [1955]). Profile of the mechanical engineer III. Personality. *Personnel Psychology, 8*(4), 469–490.

Harvey, W. B., & Anderson, E. L. (2005). *Minorities in higher education: Twenty-first annual status report 2003–2004*. Washington, DC: American Council on Education.

Holland, J. M., Major, D. A., Morganson, V., & Orvis, K. A. (2011). Capitalizing on opportunity outside the classroom: Exploring supports and barriers to the professional development activities of computer science and engineering majors. *Journal of Women and Minorities in Science and Engineering, 17*(2), 173–192.

Holtzapple, M. T., & Reece, W. D. (2003). *Foundations of engineering*. New York: McGraw-Hill.

Hug, S., & Jurow, A. S. (2013). Learning together or going it alone: How community contexts shape the identity development of minority women in computing. *Journal of Women and Minorities in Science and Engineering, 19*(4), 293–292.

Hurtado, S., Eagan, M. K., Cabrera, N. L., Lin, M. H., Park, J., & Lopez, M. (2008). Training future scientists: Predicting first year minority student participation in health science research. *Research in Higher Education, 49*(1), 126–152.

Hurtado, S., Eagan, M. K., Tran, M., Newman, C., Chang, M. J., & Velasco, P. (2011). "We do science here": Underrepresented students' interactions with faculty in different college contexts. *Journal of Social Issues, 67*(3), 553–579.

Hurtado, S., Han, J. C., Saenz, V. B., Espinosa, L. L., Cabrera, N. L., & Cerna, O. S. (2007). Predicting transition and adjustment to college: Biomedical and behavioral science aspirants' and minority students' first year of college. *Research in Higher Education, 48*(7), 841–887.

Ing, M., & Denson, N. (2013). Entering first-year students' openness to diversity: A comparison of intended engineering majors with other majors within an ethnically diverse institution. *Journal of Women and Minorities in Science and Engineering, 19*(4), 349–363.

Irvine, J. J., & Fenwick, L. T. (2011). Teachers and teaching for the new millennium: The role of HBCUs. *Journal of Negro Education, 80*(3), 197–208.

Ishitani, T. T. (2006). Studying attrition and degree completion behavior among first generation students in the United States. *Journal of Higher Education, 77*(5), 861–885.

Johnson, D. R., Soldner, M., Leonard, J. B., Alvarez, P., Inkelas, K. K., Rowan-Kenyon, H. T., & Longerbeam, S. D. (2007). Examining sense of belonging among first-year undergraduates from different racial/ethnic groups. *Journal of College Student Development, 48*(5), 525–542.

Johnson, K. V., & Watson, E. D. (2005). A historical chronology of the plight of African Americans gaining recognition in engineering and technology. *Journal of Technology Studies, 31*(2), 81–93.

Jones, M. T., Barlow, A. E. L., & Villarejo, M. (2010). Importance of undergraduate research for minority persistence and achievement in biology. *Journal of Higher Education, 81*(1), 82–115.

Jussim, L., & Harber, K. D. (2005). Teacher expectations and self-fulfilling prophecies: Known and unknowns, resolved and unresolved controversies. *Personality and Social Psychology Review, 9*(2), 131–155.

Karkouti, I. M. (2016). Black students' educational experiences in predominantly White institutions: A review of related literature. *The College Student Journal, 50*(1), 59–70.

Kim, M. M. (2002). Historically Black vs. White institutions: Academic development among Black students. *The Review of Higher Education, 25*(4), 385–407.

Kim, M. M. (2004). The experience of African-American students in historically Black institutions. *Thought & Action, Summer,* 107–124.

Kim, M. M., & Conrad, C. F. (2006). The impact of historically Black colleges and universities on the academic success of African-American students. *Research in Higher Education, 47*(4), 399–427.

Kim, D., & Otts, C. (2010). The effect of loans on time to doctorate degree: Differences by race/ethnicity, field of study, and institutional characteristics. *Journal of Higher Education, 81*(1), 1–31.

Kim, Y. K., & Sax, L. J. (2009). Student-faculty interaction in research universities: Differences by student gender, race, social class, and first- generation status. *Research in Higher Education, 50*(5), 437–459.

Kokkelenberg, E. C., & Sinha, E. (2010). Who succeeds in STEM studies? An analysis of Binghamton University undergraduate students. *Economics of Education Review, 29*(6), 935–946.

Kuh, G. (2008). *High-impact educational practices: What they are, who has access to them, and why they matter.* Washington, DC: Association of American Colleges and Universities.

Lain, M. A., & Frehill, L. M. (2012). *2010–2011 NACME graduating scholars survey results.* White Plains, NY: National Action Council for Minorities in Engineering.

Laird, T. F. N., Bridges, B. K., Morelson-Quainoo, C. L., Williams, J. M., & Holmes, M. S. (2007). African American and Hispanic student engagement at minority-serving and predominantly White institutions. *Journal of College Student Development, 48*(1), 39–56.

Lam, P. C., Doverspike, D., & Mawasha, R. P. (1997). Increasing diversity in engineering academics (IDEAs): Development of a program for improving African American representation. *Journal of Career Development, 24*(1), 55–79.

Lam, P. C., Doverspike, D., & Mawasha, R. P. (1999). Predicting success in a minority engineering program. *Journal of Engineering Education, 88*(3), 265–267.

Lam, P., Mawasha, P., & Chu, W. (1994). Impact of workshops and permanent study center on the minority engineering tutor incentive program at The University of Akron. *Proceedings of the ASEE Gulf-Southwest,* 465–475.

Lam, P. C., Srivatsan, T., Doverspike, D., Vesalo, J., & Mawasha, P. R. (2005). A ten year assessment of the pre-engineering program for under-represented, low income and/or first generation college students at the University of Akron. *Journal of STEM Education: Innovations and Research, 6*(3/4), 14–19.

Lam, P. C., Ugweje, O., Mawasha, P. R., & Srivatsan, T. S. (2003). An assessment of the effectiveness of the McNair program at the University of Akron. *Journal of Women and Minorities in Science and Engineering, 9*(1),103–117.

Landis, R. (2005). *Retention by design.* New York: National Action Council for Minorities in Engineering.

Lee, W. C., Brozina, C., Amelink, C. T., & Jones, B. D. (2017). Motivating incoming engineering students with diverse backgrounds: Assessing a summer bridge program's impact on academic motivation. *Journal of Women and Minorities in Science and Engineering, 23*(2), 121–145.

Lee, W. C., Matusovich, H. M., & Brown, P. R. (2014). Measuring underrepresented student perceptions of inclusions within engineering departments and universities. *International Journal of Engineering Education, 30*(1), 150–165.

Lee, J. M., & Rawls, A. (2010). *The college completion agenda: 2010 progress report.* New York: The College Board. Retrieved from http://completionagenda.college board.org/sites/default/files/reports_pdf.Progress_Report_2010.pdf.

Lewis, B. F. (2003). A critique of the literature on the underrepresentation of African Americans in science: Directions for future research. *Journal of Women and Minorities in Science and Engineering, 9*(3/4), 361–373.

Lewis, J. A., Mendenhall, R., Harwood, S. A., & Hunt, M. B. (2012). Coping with gendered racial microaggressions among Black women college students. *Journal of African American Studies, 17*(1), 51–73.

Leyva, L. A. (2016). An intersectional analysis of Latina college women's counter-stories in mathematics. *Journal of Urban Mathematics Education, 9*(2), 81–121.

Lord, S. M., Camacho, M. M., Layton, R. A., Long, R. A., Ohland, M. W., & Wasburn, M. H. (2009). Who's persisting in engineering? A comparative analysis of female and male Asian, black, Hispanic, Native American, and white students. *Journal of Women and Minorities in Science and Engineering, 15*(2), 167–190.

Lord, S., Layton, R., Ohland, M., Brawner, K., & Long, R. (2014). A multi-institution study of student demographics and outcomes in chemical engineering. *Chemical Engineering Education, 48*(4), 231–238.

Lord, S. M., Layton, R. A., Ohland, M. W., & Orr, M. K. (2013, October). Student demographics and outcomes in electrical and mechanical engineering. In *Frontiers in Education Conference*, Oklahoma City, OK, 57–63. New York: IEEE.

Lutkus, A. D., Lauko, M., & Brockway, D. (2006). *The nation's report card: Trial urban district assessment of grades 4 and 8: Science 2005.* Washington, DC: U.S. Department of Education, National Assessment of Educational Progress.

Malcolm-Piquex, L. E., & Malcolm, S. (2015). African American women and men into engineering: Are some pathways smoother than others? In J. B. Slaughter, Y. Tao, & W. Pearson, Jr. (Eds.), *Changing the face of engineering: The African American experience* (pp. 90–119). Baltimore, MD: Johns Hopkins University Press.

Margolis, J., & Fisher, A. (2002). *Unlocking the clubhouse: Women in computing.* Cambridge, MA: MIT Press.

Marra, R. M., Rodgers, K. A., & Shen, D. (2012). Leaving engineering: A multi-year single institution study. *Journal of Engineering Education, 101*(1), 6–27.

Martin, J. P., Simmons, D. R., & Yu, S. L. (2013). The role of social capital in the experiences of Hispanic women engineering majors. *Journal of Engineering Education, 102*(2), 227–243.

Maton, K. I., Domingo, M. R. S., Stolle-McAllister, K. E., Zimmerman, J. L., & Hrabowski, F. A. III. (2009). Enhancing the number of African Americans who pursue STEM Ph.Ds: Meyerhoff scholarship program outcomes, processes, and individual predictors. *Journal of Women and Minorities in Science and Engineering, 15*(1), 15–37.

Maton, K. I., & Hrabowski III, F. A. (2004). Increasing the number of African American PhDs in the sciences and engineering: A strengths-based approach. *American Psychologist, 59*(6), 547–556.

Maton, K. I., Hrabowski, F. A. III, & Schmitt, C. L. (2000). African American college students excelling in the sciences: College and post-college outcomes in the Meyerhoff Scholars Program. *Journal of Research in Science Teaching, 37*(7), 629–654.

Maton, K. I., Watkins-Lewis, K. M., Beason, T., & Hrabowski, F. A. III. (2015). Enhancing the number of African Americans pursuing the Ph.D. in engineering: Outcomes and processes in the Meyerhoff Scholarship Program. In J. B. Slaughter, Y. Tao, & W. Pearson, Jr. (Eds.), *Changing the face of engineering: The African American experience* (pp. 354–386). Baltimore, MD: Johns Hopkins University Press.

McPhail, I. P. (2015). Enhancing the community college pathway to engineering careers for African American students: A critical review of promising and best practices. In J. B. Slaughter, Y. Tao, & W. Pearson, Jr. (Eds.), *Changing the face of engineering: The African American experience* (pp. 305–334). Baltimore, MD: Johns Hopkins University Press.

Meyer, M., & Marx, S. (2014). Engineering dropouts: A qualitative examination of why undergraduates leave engineering. *Journal of Engineering Education, 103*(4), 525–548.

Millett, C. M., & Nettles, M. T. (2006). Expanding and cultivating the Hispanic STEM doctoral workforce: Research on doctoral student experiences. *Journal of Hispanic Higher Education, 5*(3), 258–287.

Miner, K., Diaz, I., & Rinn, A. (2017). Incivility, psychological distress, and math self-concept among women and students of color in STEM. *Journal of Women and Minorities in Science and Engineering, 23*(3), 211–230.

Morning, C., & Fleming, J. (1994). Project preserve: A program to retain minorities in engineering. *Journal of Engineering Education, 83*(3), 237–242.

Morning, C., & Fleming, J. (2012). Doubling the number of minority students in science, engineering and math. In J. Fleming, *Enhancing the academic performance and retention of minorities: What we can learn from program evaluation* (pp. 135–154). San Francisco: Jossey-Bass.

NAE (National Academy of Engineering). (2008). *Changing the conversation: Messages for improving public understanding of engineering.* Washington, DC: National Academies Press. Retrieved from www.nap.edu/catalog.php?record_id=12187.

NCES (National Center for Education Statistics). (2013). *Digest of education statistics 2013 (NCES 2014–2015).* Washington, DC: Author. Retrieved from http://nces.ed.gov/programs/coe/index.asp.

NACME (National Action Council for Minorities in Engineering). (2008). *The NACME/Qualcom community college pre-engineering studies transfer scholarship program: An analysis of the current efforts.* White Plains, NY: Author.

NACME (National Action Council for Minorities in Engineering). (2010). Community college transfers and engineering bachelor's degree programs. *Research & Policy Brief, 1*(1, September), 1–2.

NACME (National Action Council for Minorities in Engineering). (2011). Latinos in engineering. *Research & Policy Brief*, 1(7, September), 1–2.

NACME (National Action Council for Minorities in Engineering). (2012). African Americans in engineering. *Research & Policy Brief*, 2(4, September), 1–2.

NACME (National Action Council for Minorities in Engineering). (2014a). African Americans in engineering. *Research & Policy Brief*, 4(1, April), 1–2.

NACME (National Action Council for Minorities in Engineering). (2014b). Latinos in engineering. *Research & Policy Brief*, 4(3, October), 1–2.

NACME (National Action Council for Minorities in Engineering). (2014c). American Indians/Alaska Natives in engineering. *Research & Policy Brief*, 4(2, August), 1–2.

Nunez, A. M., Hurtado, S., & Galdeano, E. C. (2015). Why study Hispanic-serving institutions? In A. M. Nunez, S. Hurtado, & E. C. Galdeano (Eds.), *Hispanic-serving institutions: Advancing research and transformative practice* (pp. 1–22) New York: Routledge.

Ohland, M. W., Brawner, C. E., Camacho, M. M., Layton, R. A., Long, R. A., Lord, S. M., & Wasburn, M. H. (2011). Race, gender, and measures of success in engineering education. *Journal of Engineering Education*, 100(2), 225–252.

Ohland, M. W., Lord, S. M., & Layton, R. A. (2015). Student demographics and outcomes in civil engineering in the United States. *Journal of Professional Issues in Engineering Education and Practice*, 141(4), 04015003, 1–7.

Oliver, M. L., Rodriguez, C. J., & Mickelson, R. A. (1985). Brown and black in white: The social adjustment and academic performance of Chicano and Black students in a predominately White university. *The Urban Review*, 17(1), 3–24.

Ong, M., Wright, C., Espinosa, L., & Orfield, G. (2011). Inside the double bind: A synthesis of empirical research on undergraduate and graduate women of color in science, technology, engineering, and mathematics. *Harvard Educational Review*, 81(2), 172–209.

Orr, M. K., Lord, S. M., Layton, R. A., & Ohland, M. W. (2014). Student demographics and outcomes in mechanical engineering in the US. *International Journal of Mechanical Engineering Education*, 42(1), 48–60.

Palmer, R. T., Davis, R. J., & Maramba, D. C. (2010). Role of an HBCU in supporting academic success for underprepared black males. *The Negro Educational Review*, 61(1–4), 85–10.

Palmer, R. T., & Young, E. M. (2009). Determined to succeed: Salient factors that foster academic success for academically unprepared black males at a black college. *Journal of College Student Retention*, 10(4), 465–482.

Pascarella, E. T., & Terenzini, P. T. (2005). *How college affects students*. San Francisco: Jossey-Bass.

Pearson, W. (2005). *Beyond small numbers: Voices of African American PhD chemists*. New York: Elsevier Science.

Pearson, W., & Marshall, E. (2016). *Kentucky-West Virginia LSAMP program evaluation*. Unpublished manuscript. Lexington, KY: University of Kentucky.

Pearson Jr, W., & Miller, J. D. (2012). Pathways to an engineering career. *Peabody Journal of Education*, 87(1), 46–61.

Perna, L., Lundy-Wagner, V., Drezner, N. D., Gasman, M., Yoon, S., Bose, E., & Gary, S. (2009). The contribution of HBCUs to the preparation of African American women for STEM careers: A case study. *Research in Higher Education*, 50(1), 1–23.

Person, D. R., & Fleming, J. (2012). Who will do math, science, engineering and technology? Academic achievement among minority students in seventeen institutions. In J. Fleming, *Enhancing the retention and academic performance of minorities: What we can learn from program evaluation* (pp. 120–134). San Francisco: Jossey-Bass.

Pierre, P. A. (2015). A brief history of the collaborative minority engineering effort: A personal account. In J. B. Slaughter, Y. Tao, & W. Pearson, Jr. (Eds.), *Changing the face of engineering: The African American experience* (pp. 13–36). Baltimore, MD: Johns Hopkins University Press.

Plett, M., Lane, A., & Peter, D. M. (2016). *Understanding diverse and atypical engineering students: Lessons learned from community college transfer scholarship recipients*. Proceedings of the 2016 Annual Conference of the American Association of Engineering Education, New Orleans, Louisiana. Retrieved from https://peer.asee.org/14483.

Powell, J. H. (2006). Engineering education. In B. A. Osif (Ed.), *Using the engineering literature* (pp. 269–285). New York: Routledge.

Prescott, B. T., & Bransberger, P. (2012). *Knocking at the college door: Projections of high school graduates* (8th ed.). Boulder, CO: Western Interstate Commission for Higher Education.

Ransom, T. (2015). Clarifying the contributions of historically black colleges and universities in engineering education. In J. B. Slaughter, Y. Tao, & W. Pearson, Jr. (Eds.), *Changing the face of engineering: The African American experience* (pp. 120–148). Baltimore, MD: Johns Hopkins University Press.

Reichert, M., & Absher, M. (1997). Taking another look at educating African American engineers: The importance of undergraduate retention. *Journal of Engineering Education, 86*(3), 241–253.

Ro, H. K., Knight, D. B., & Loya, K. I. (2016). Exploring the moderating effects of race and ethnicity on the relationship between curricular and classroom experiences and learning outcomes in engineering. *Journal of Women and Minorities in Science and Engineering, 22*(2), 91–118.

Ro, H. K., & Loya, K. I. (2015). The effect of gender and race intersectionality on student learning outcomes in engineering. *The Review of Higher Education, 38*(3), 359–396.

Rodriguez, A., & Galdeano, E. C. (2015). Do Hispanic-serving institutions really underperform? Using propensity score matching to compare outcomes of Hispanic serving and Non-Hispanic-serving institutions. In A. M. Nunez, S. Hurtado, & E. C. Galdeano (Eds.), *Hispanic-serving institutions: Advancing research and transformative practice* (pp. 196–216). New York: Routledge.

Rosati, P. (2003). *Student performance in chosen engineering discipline related to personality type*. Unpublished manuscript, Department of Civil Engineering, University of Western Ontario. Retrieved from www.siu.edu/~coalctr/paper301.htm.

Rosser, S. V. (2012). *Breaking into the lab: Engineering progress for women in science*. New York: New York University Press.

Rovai, A. P., Gallien, L. B., & Wighting, M. J. (2005). Cultural and interpersonal factors affecting African American academic performance in higher education. *Journal of Negro Education, 74*(4), 359–370.

Ryu, M. (2010). *Minorities in higher education: 24th status report*. Washington, DC: American Council on Education.

Sax, L. J., Zimmerman, H. B., Blaney, J. M., Toven-Lindsey, B., & Lehman, K. J. (2017). Diversifying undergraduate computer science: The role of department chairs in promoting gender and racial diversity. *Journal of Women and Minorities in Science and Engineering*, 23(2), 101–119.

Scissons, E. H. (1979). Profiles of ability: Characteristics of Canadian engineers. *Engineering Education*, 69(8), 822–826.

Scissons, E. H. (1984). Profiles of ability: The engineer revisited. *Engineering Education*, 75(3), 165–168.

Seifert, T. A., Drummond, J., & Pascarella, E. T. (2006). African American students' experiences of good practices: A comparison of institutional type. *Journal of College Student Development*, 47(2), 185–205.

Seymour, E., & Hewitt, N. (1997). *Talking about leaving*. Boulder, CO: Westview Press.

Shehab, R., Murphy, T. J., & Foor, C. E. (2012). Do they even have that anymore? The impact of redesigning a minority engineering program. *Journal of Women and Minorities in Science and Engineering*, 18(3), 235–253.

Shorette, C. R., & Palmer, R. T. (2015). Historically Black Colleges and Universities (HBCUs): Critical facilitators of non-cognitive skills for Black males. *Western Journal of Black Studies*, 39(1), 18–29.

Simon, R. M., & Farkas, G. (2008). Sex, class, and physical science educational attainment: Portions due to achievement versus recruitment. *Journal of Women and Minorities in Science and Engineering*, 14(3), 269–300.

Slaton, A. (2011). *Race, rigor, and selectivity in U.S. engineering: The history of an occupational color line*. Cambridge, MA: Harvard University Press.

Smith, S. S., & Moore, M. R. (2002). Expectations of campus racial climate and social adjustment among African American college students. In W. R. Allen, M. B. Spencer, & C. O'Connor (Eds.), *African American education: Race community, inequality and achievement* (pp. 93–118). Burlington, MA: Elsevier Science.

Smith, W. A. (2009). Campuswide climate: Implications for African American students. In L. C. Tillman (Ed.), *The SAGE handbook of African American education* (pp. 297–309). Thousand Oaks, CA: Sage.

Solberg, V. S., & Viliarreal, P. (1997). Examination of self-efficacy, social support, and stress as predictors of psychological and physical distress among Hispanic college students. *Hispanic Journal of Behavioral Sciences*, 19(2), 182–201.

Solórzano, D., Ceja, M., & Yosso, T. (2000). Critical race theory, racial micro-aggressions, and campus racial climate: The experiences of African American college students. *Journal of Negro Education*, 69(1/2), 60–73.

Sondgeroth, M. S., & Stough, L. M. (1992). *Factors influencing the persistence of ethnic minority students enrolled in a college engineering program*. Paper presented American Education Research Association, San Francisco, CA. ERIC Document No. ED 353 923.

Stewart, A. J., Malley, J. E., & Herzog, K. A. (2016). Increasing the representation of women faculty in STEM departments: What makes a difference? *Journal of Women and Minorities in Science and Engineering*, 22(1), 23–47.

Su, L. K. (2010). Quantification of diversity in engineering higher education in the United States. *Journal of Women and Minorities in Science and Engineering*, 16(2), 161–175.

Suarez-Balcazar, Y., Orellana-Damacela, L., Portillo, N., Rowan, J. M., & Andrews-Guillen, C. (2003). Experiences of differential treatment among college students of color. *The Journal of Higher Education, 74*(4), 428–444.

Suresh, R. (2006/2007). The relationship between barrier courses and persistence in engineering. *Journal of College Student Retention, 8*(2), 215–239.

Swail, W. S., & Chubin, D. E. (2007). *An evaluation of the NACME block grant program.* Virginia Beach, VA: Educational Policy Institute.

Tennenbaum, H. R., & Ruck, M. D. (2007). Are teacher's expectations different for racial minority than for European American students? A meta-analysis. *Journal of Educational Psychology, 99*(2), 253–273.

Tang, J. (2000). *Doing engineering: The career attainment and mobility of Caucasian, Black and Asian American engineers.* Lanham, MD: Row and Littlefield.

Tomlinson-Clarke, S. (1998). Dimensions of adjustment among college women. *Journal of College Student Development, 39*(4), 364–372.

Tsui, L. (2007). Effective strategies to increase diversity in STEM fields: A review of the research literature. *Journal of Negro Education, 76*(4), 555–581.

U.S. Census Bureau. (2013). *2012 national population projections.* Washington, DC: Author. Retrieved from www.census.gov/population/projections/data/national/2012.html.

Vogt, C. M. (2008). Faculty as a critical juncture in student retention and performance in engineering programs. *Journal of Engineering Education, 97*(1), 27–36.

Walden, S. E., & Foor, C. (2008). What's to keep you from dropping out? *Journal of Engineering Education, 97*(2), 191–206.

Wao, H. O., Lee, R. S., & Borman, K. M. (2010). Climate for retention to graduation: A mixed methods investigation of student perceptions of engineering departments and programs. *Journal of Women and Minorities in Science and Engineering, 16*(4), 293–318.

Wharton, D. E. (1992). *A struggle worthy of note: The engineering and technological education of black Americans.* Westport, CT: Greenwood Press.

Williams, T. (2017). America's top CEOs and their college degrees. *Investopedia,* August 31.

Wilson-Lopez, A., Mejia, J. A., Hasbún, I. M., & Kasun, G. S. (2016). Latina/o adolescents' funds of knowledge related to engineering. *Journal of Engineering Education, 105*(2), 278–311.

Wolfe, J., Powell, B. A., Schlisserman, S., & Kirshon, A. (2016). *Teamwork in engineering undergraduate classes: What problems do students experience?* Proceedings of the 123rd Annual Meeting of the American Association of Engineering Education(ASEE), New Orleans, LA. ID Number 16447.

Wolf-Wendel, L. E. (2000). Women friendly campuses: What five institutions are doing right? *Review of Higher Education, 23*(3), 319–345.

Yelamarthi, K., & Mawasha, P. R. (2008). A pre-engineering program for the under-represented, low-income and/or first-generation college students to pursue higher education. *Journal of STEM Education: Innovations and Research, 9*(3/4), 5–15.

Yoder, B. (2012). *Going the distance in engineering education: Best practices and strategies for retaining engineering, engineering technology, and computing students.* Washington, DC: American Association for Engineering Education (ASEE).

Zimmerman, A. P., Johnson, R. G., Hoover, T. S., Hilton, J. W., Heinemann, P. H., & Buckmaster, D. R. (2006). Comparison of personality types and learning styles of engineering students, agricultural systems management students, and faculty in an agricultural and biological engineering department. *Transactions of the American Society of Agricultural Biological Engineers (ASABE)*, 49(1), 311–317.

Zywno, M. S., & Waalen, J. K. (2002). The effect of individual learning styles on student outcomes in technology-enabled education. *Global Journal of Engineering Education, Australia*, 6(1), 35–44.

3 Performance and Retention to Graduation Rates of Minority Students in NACME Block Grant Institutions
Analysis of Aggregate Statistical Data

Summary

Aggregate statistical data was gathered from 26 NACME consortium member engineering schools as a first step in the search for factors that might facilitate the success of underrepresented minority students in engineering. The purpose of this analysis was threefold: (1) to provide an institutional context for th studies of student focus groups and student surveys to follow, (2) to determine any significant differences between minorities and non-minorities in raw (i.e., unadjusted) academic performance and retention to graduation rates, and (3) to determine what differences remained after adjusting academic performance and retention for differences in test scores. Analysis of variance was used to determine significant ethnic differences in average performance, while linear regression procedures were used to determine institutional rankings before and after variation due to standardized test scores were removed.

The results showed that with test scores controlled, underrepresented minority students as a group showed no differences in math GPA, GPA in core engineering courses, 1-year retention, 4-year, or 6-year graduation rates compared to non-minority students. Minority students did exhibit lower overall GPA and GPA in basic science courses. Within each ethnic group, females exhibited better overall GPA compared to their male counterparts, while non-minority females also showed higher 6-year graduation rates. Many of the differences typically observed between minority and non-minority students in graduation rates may be due to initial differences in entering standardized test scores. However, some evidence of academic underperformance was found among minorities, even with test scores equal to non-minorities. The possible roles of systemic educational issues as well as adjustment issues were discussed.

Performance and Retention to Graduation Rates of Minority Students in NACME Block Grant Institutions: Analysis of Aggregate Statistical Data

The foregoing review of the literature and overview of the study have established a number of contextual considerations for minorities in engineering. First, it is in the national interest to facilitate the development of homegrown underrepresented minorities in engineering; i.e., African American, Hispanic, and Native American students (Chubin, Donaldson, Olds, & Fleming, 2008; Chubin, May, & Babco, 2005; Slaughter, 2015). Second, a sparse literature focuses exclusively on underrepresented minorities in engineering; engineering students are typically included in studies of minorities in STEM fields, and this co-mingling of students in various sciences may obscure important trends (Atwaters, Leonard, & Pearson, 2015). Third, few comprehensive studies of minorities in engineering exist, and greater quantification of factors that might facilitate programmatic and student success in this area are needed (Slaton, 2011; Swail & Chubin, 2007). Fourth, previous studies tend to find more similarities than differences in the experiences of African American and Hispanic college students in Predominantly White Institutions (Cole, 2008; Constantine, Wilton, & Caldwell, 2003; Ishitani, 2006; Lee & Rawls, 2010), if not in their respective Minority-Serving Institutions (Bridges, Kinzie, Laird, & Kuh, 2008; Gasman, 2008), but further study of their respective experiences in engineering are needed.

This three-pronged study of success factors for minorities in engineering begins with an analysis of institutional statistics that might provide context for the focus group and student survey studies to follow. The overall goal was to unearth differences or similarities between minorities and non-minorities that shed light on factors that facilitate the success of engineering students. NACME consortium member institutions have records of strong Minority in Engineering Programs (MEPs) with retention to graduation rates that may equal or exceed the national average. Therefore, it seems important to document these rates to provide baselines for the succeeding studies. At issue are differences between minority and non-minority students, and any differences among ethnic groups in these rates. Less attention has been paid in the literature to differences between groups in course grades, but an examination of them may provide further insight into factors facilitating the success of students in engineering. Several studies have attempted to rank the high producers of minorities in engineering (Malcolm-Piquex & Malcolm, 2015; Ransom, 2015), and so, this study engaged in a similar ranking of NACME institutions participating in this study.

Previous studies have reported that dropout rates from engineering in general are high—around 50%—but are higher still for minorities and have remained higher over time. Landis (2005) reported a 67% dropout rate. Borrego, Padilla, Zhang, Ohland, and Anderson (2005) found dropout rates of 46%–73% for minorities, compared to 34%–60% for non-minorities, at nine southeastern universities including two HBCUs. Yoder (2012) reported dropout rates of 61.3% for African Americans, 55.6% for Hispanics, and 61.4% for Native Americans, compared to 33.5% for Asian Americans and 40.3% for Caucasians. NACME (2012) reported a dropout rate of 68.8% for African Americans. Studies that compare academic performances generally found that minorities exhibited poorer grades, with or without better test scores (Araque, Roldan, & Salguero, 2009; Borrego et al., 2005; Chen & Weko, 2009; Suresh, 2006/2007). According to Marra, Rodgers, and Shen (2012), minority students more often cited academic reasons for leaving engineering compared to non-minority students, and Borrego et al. (2005) found lower GPAs among minorities who leave engineering. There is evidence that women in engineering achieve higher grade averages and have higher graduation rates than their male counterparts (Borrego et al., 2005; Brawner, Camacho, Lord, Long, & Ohland, 2012; Lord et al., 2009; Orr, Lord, Layton, & Ohland, 2014). Adelman (1998), however, found that women were less likely to graduate. Black and Latinas have also shown a pattern of better performance and/or and higher graduation rates in engineering than their male counterparts (Ohland et al., 2011), despite the discomfort of a chilly climate (Camacho & Lord, 2013). It has been well-documented that HBCUs graduate a disproportionate share of engineering graduates (Chubin et al., 2005; Malcolm-Piquex & Malcolm, 2015; Ransom, 2015), but no similar advantages for Hispanic students in Hispanic-serving engineering institutions have yet been reported. This chapter seeks to determine the extent to which the above trends can be found in NACME member consortium engineering schools.

One unique feature of this investigation was the institution of a control for differences in student abilities measured by standardized test scores because we were interested in comparing students of roughly similar abilities. Depending on the analysis, several statistical methods were used to equalize outcomes on this basis that are described in the Method and Appendix 3.1. The present study investigated both the aggregate academic performance and retention to graduation of minorities in a sample of 26 NACME consortium engineering schools and compared their academic performance to non-minorities. The study was guided by the following research questions:

1. Are there significant differences between minorities and non-minorities in raw academic performance in the NACME consortium institutions, and are there variations depending on the courses examined? What differences remain after adjusting academic performance for differences in test scores?
2. Are there significant differences in the raw minority and non-minority retention to graduation rates at 1-year, 4-year and 6-year intervals? What differences remain after adjustments for test scores?
3. Which schools in this consortium have the highest retention to graduation rates for minorities in engineering? The issue here is whether

schools with the better retention to graduation rates are those with the best students (i.e., in terms of SAT/ACT scores), or those with the best programs?

4. Are there significant ethnic and gender differences among minority and non-minority students in the variables under investigation?

5. Do the results shed light on the reasons for any minority and non-minority differences that may be linked to prior systemic educational issues, or to issues of institutional adjustment?

Method

All 31 NACME member institutions were potential participants in the study. However, the final sample included 26 institutions for a variety of reasons including an insufficient length of time in the consortium to permit analyses of 6-year graduation rates. Participating institutions were all four-year institutions representing 18 states. The number included two HBCUs and six Hispanic-Serving Institutions (HSIs).

Each institution was asked to provide data on total enrollment, a series of GPA measures, and retention to graduation rates for the base year of 2012. Data was requested for non-minorities as a whole, as well as male and female non-minorities, minorities as a group, African American males and females (i.e., separately by gender), Hispanic males and females, and Native American males and females. The numbers of Native American students were miniscule and did not permit separate analysis. No ethnic breakdown was requested for non-minorities. Because African American, Hispanic, and Native American/Pacific Islander data was requested separately by gender, several institutions were unable to report data for small numbers of students. Therefore, the number of institutions reporting data for minorities as a whole was greater than the number reporting male and female data within minority ethnic groups. Details of the method can be found in Appendix 3.1.

Results

Academic Performance

If there is a sparse literature that focuses on minorities in engineering, fewer still are studies that compare the engineering grades of underrepresented minorities with non-minorities. The present study afforded an opportunity to compare the groups of students on core engineering course grades of math, science, and engineering as well as overall GPA for the 2012 cohort across the 26 participating institutions.

The math and verbal test scores of these two groups of students were significantly different, with the minority average SAT (or SAT equivalent) lower than that of non-minorities (1,124.8 vs. 1,210.7). Table 3.1 and Figure 3.1 show that before controlling for test scores, there were

Table 3.1 Academic Performance of Minorities and Non-Minorities by Ethnicity in NACME Block Grant Institutions

Measure	Raw Academic Performance			Test Score-Adjusted Academic Performance		
	Minorities	Non-Minorities	Significance	Minorities	Non-Minorities	Adjusted Significance
SAT	1124.8 SD = 114.7 n = 26	1210.7 SD = 103.3 n = 24	$F = 7.71^{**}$ Eta = .138	NA	NA	NA
Overall GPA	2.80 (.033) n = 26	2.99 (.034) n = 24	$F = 16.16^{***}$ Eta = .252	2.84 (.027)	2.95 (.029)	$F = 7.28^{**}$ Eta = .134 $F_{(SATM)} = 6.19^{*}$ $F_{(SATV)} - 10.02^{**}$
Math GPA	2.55 (.062) n = 24	2.76 (.063) n = 23	$F = 5.32^{*}$ Eta = .106	2.61 (.053)	2.69 (.054)	$F = 1.04$ ns $F_{(SATM)} = 0.73$ ns $F_{(SATV)} = 1.63$ ns
Science GPA	2.56 (.061) n = 24	2.87 (.062) n = 23	$F = 12.05^{***}$ Eta = .211	2.62 (.054)	2.81 (.056)	$F = 5.72^{*}$ Eta = .115 $F_{(SATM)} = 4.30^{*}$ $F_{(SATV)} = 6.77^{*}$
Engineering GPA	2.99 (.077) n = 20	3.11 (.079) n = 19	$F = 1.20$ ns	3.04 (.073)	3.06 (.075)	$F = .013$ ns $F_{(SATM)} = .011$ ns $F_{(SATV)} = .012$ ns

(Continued)

Table 3.1 (Continued)

| Measure | Raw Academic Performance | | | Test Score-Adjusted Academic Performance | | |
	Minorities	Non-Minorities	Significance	Minorities	Non-Minorities	Adjusted Significance
1-Year Retention	77.1% (2.41) n = 24	81.0% (2.52) n = 22	$F = 1.27$ ns	80.0% (2.07)	77.8% (2.17)	$F = 0.49$ ns $F_{(SATM)} = 0.74$ ns $F_{(SATV)} = 0.25$ ns
4-Year Graduation	13.6% (2.58) n = 24	23.3% (2.69) n = 22	$F = 6.89^{*}$ $Eta = .135$	17.1% (2.02)	19.5% (2.12)	$F = 0.58$ ns $F_{(SATM)} = 0.37$ ns $F_{(SATV)} = 1.59$ ns
6-Year Graduation	37.3% (3.33) n = 25	49.3% (3.39) n = 24	$F = 6.33^{*}$ $Eta = .119$	40.5% (2.98)	46.0% (3.04)	$F = 1.59$ ns $F_{(SATM)} = 1.03$ ns $F_{(SATV)} = 1.66$ ns

(Continued)

Table 3.1 (Continued)

Measure	Raw Academic Performance			Test Score-Adjusted Academic Performance		
	African Americans	Non-Minorities	Significance	African Americans	Non-Minorities	Adjusted Significance
SAT	1106.7 (16.1) n = 46	1208.6 (15.7) n = 48	$F = 20.56^{***}$ Eta = .183	NA	NA	NA
Overall GPA	2.75 (.037) n = 46	3.05 (.036) n = 48	$F = 33.75^{***}$ Eta = .268	2.80 (.035)	2.99 (.034)	$F = 14.69^{***}$ Eta = .139 $F_{(SATM)} = 13.15^{***}$ $F_{(SATV)} = 17.34^{***}$
Math GPA	2.55 (.069) n = 42	2.79 (.066) n = 46	$F = 6.57^{*}$ Eta = .071	2.63 (.068)	2.73 (.065)	$F = 1.04$ ns $F_{(SATM)} = 0.61$ ns $F_{(SATV)} = 1.28$ ns
Science GPA	2.46 (0.65) n = 39	2.87 (.060) n = 46	$F = 22.21^{***}$ Eta = .211	2.52 (.064)	2.82 (.059)	$F = 10.99^{***}$ Eta = .118 $F_{(SATM)} = 9.26^{**}$ $F_{(SATV)} = 11.42^{***}$
Engineering GPA	2.92 (.097) n = 34	3.14 (.092) n = 38	$F = 2.73$ ns	3.01 (.100)	3.05 (.094)	$F = 0.11$ ns $F_{(SATM)} = 0.04$ ns $F_{(SATV)} = 0.31$ ns

(Continued)

Table 3.1 (Continued)

Measure	Raw Academic Performance			Test Score-Adjusted Academic Performance		
	African Americans	Non-Minorities	Significance	African Americans	Non-Minorities	Adjusted Significance
1-Year Retention	76.8% (2.51) n = 38	83.3% (2.33) n = 44	$F = 2.59$ ns	79.9% (2.51)	79.6% (2.31)	$F = 0.01$ ns $F_{(SATM)} = 0.29$ ns $F_{(SATV)} = 0.02$ ns
4-Year Graduation	12.3% (2.44) n = 35	25.1% (2.20) n = 43	$F = 15.34^{***}$ $Eta = .168$	16.1% (2.29)	22.0% (2.04)	$F = 3.36, p < .10$ $F_{(SATM)} = 2.91$ ns $F_{(SATV)} = 6.21^{*}$
6-Year Graduation	32.9% (3.25) n = 43	52.5% (3.07) n = 48	$F = 19.28^{***}$ $Eta = .178$	35.6% (3.31)	50.1% (3.11)	$F = 9.21^{**}$ $Eta = .095$ $F_{(SATM)} = 7.37^{**}$ $F_{(SATV)} = 9.51^{**}$

(Continued)

Table 3.1 (Continued)

	Raw Academic Performance			Test Score-Adjusted Academic Performance		
Measure	Hispanics n = 24	Non-Minorities n = 24	Significance	Hispanics	Non-Minorities	Adjusted Significance
SAT	1147.8 (15.7) n = 48	1208.6 (15.7) n = 48	$F = 7.55^{**}$ Eta = .074	NA	NA	NA
Overall GPA	2.85 (.031) n = 48	3.05 (.031) n = 48	$F = 20.72^{***}$ Eta = .181	2.88 (.027)	3.02 (.027)	$F = 12.38^{***}$ Eta = .117 $F_{(SATM)} = 11.69^{***}$ $F_{(SATV)} = 16.49^{***}$
Math GPA	2.54 (.060) n = 46	2.79 (.060) n = 46	$F = 9.10^{**}$ Eta = .092	2.59 (.056)	2.75 (.056)	$F = 4.36^{*}$ Eta = .047 $F_{(SATM)} = 3.74{\sim}$ $F_{(SATV)} = 5.74^{*}$
Science GPA	2.58 (.054) n = 45	2.87 (.053) n = 46	$F = 14.47^{***}$ Eta = .140	2.63 (.049)	2.82 (.049)	$F = 7.44^{**}$ Eta = .078 $F_{(SATM)} = 6.19^{*}$ $F_{(SATV)} = 9.98^{**}$
Engineering GPA	3.06 (.074) n = 36	3.14 (.072) n = 38	$F = 0.56$ ns	3.09 (.071)	3.10 (.069)	$F = 0.00$ ns $F_{(SATM)} = 0.01$ $F_{(SATV)} = 0.16$

(Continued)

Table 3.1 (Continued)

Measure	Raw Academic Performance			Test Score-Adjusted Academic Performance		
	Hispanics	Non-Minorities	Significance	Hispanics	Non-Minorities	Adjusted Significance
1-Year Retention	82.7% (1.63) n = 44	82.3% (1.63) n = 44	$F = 0.03$ ns	83.8% (1.58)	81.2% (1.58)	$F = 1.32$ ns $F_{(SATM)} = 1.49$ ns $F_{(SATV)} = 0.92$ ns
4-Year Graduation	14.9% (2.31) n = 41	25.1% (2.26) n = 43	$F = 9.98^{**}$ $Eta = .109$	17.6% (1.99)	22.5% (1.94)	$F = 2.94$ ns $F_{(SATM)} = 2.79$ ns $F_{(SATV)} = 5.29^{*}$ $Eta = .064$
6-Year Graduation	43.9% (3.62) n = 43	52.5% (3.42) n = 48	$F = 2.98$ ns	45.7% (3.59)	50.9% (3.39)	$F = 1.05$ ns $F_{(SATM)} = 1.19$ ns $F_{(SATV)} = 0.99$ ns

(Continued)

Table 3.1 (Continued)

Measure	Raw Academic Performance			Test Score-Adjusted Academic Performance		
	African Americans	Hispanics	Significance	African Americans	Hispanics	Adjusted Significance
SAT	1106.7 (17.0) n = 46	1147.8 (16.6) n = 48	$F = 2.98$ ns	NA	NA	NA
Overall GPA	2.75 (.040) n = 46	2.85 (.039) n = 48	$F = 3.10$ ns	2.77 (.038)	2.83 (.037)	$F = 1.34$ ns $F_{(SATM)} = 0.86$ ns $F_{(SATV)} = 0.82$ ns
Math GPA	2.55 (.079) n = 42	2.54 (.076) n = 46	$F = 0.01$ ns	2.58 (.077)	2.52 (.074)	$F = 0.29$ ns $F_{(SATM)} = 0.44$ ns $F_{(SATV)} = 0.35$ ns
Science GPA	2.46 (.066) n = 39	2.58 (.061) n = 45	$F = 2.02$ ns	2.48 (.062)	2.57 (.058)	$F = 1.15$ ns $F_{(SATM)} = 0.70$ ns $F_{(SATV)} = 0.65$ ns
Engineering GPA	2.92 (.098) n = 34	3.06 (.095) n = 36	$F = 1.13$ ns	2.95 (.097)	3.02 (.094)	$F = 0.24$ ns $F_{(SATM)} = 0.16$ ns $F_{(SATV)} = 0.47$ ns

(Continued)

Table 3.1 (Continued)

Measure	Raw Academic Performance			Test Score-Adjusted Academic Performance		
	African Americans	Hispanics	Significance	African Americans	Hispanics	Adjusted Significance
1-Year Retention	76.8% (2.59) n = 38	72.7% (2.42) n = 44	F = 2.75 ns	77.8% (2.52)	81.8% (2.34)	F = 1.29 ns $F_{(SATM)}$ = 0.75 ns $F_{(SATV)}$ = 1.26 ns
4-Year Graduation	12.3% (2.43) n = 35	14.9% (2.25) n = 41	F = 0.65 ns	13.3% (2.21)	14.0% (2.04)	F = 0.05 ns $F_{(SATM)}$ = 0.01 ns $F_{(SATV)}$ = 0.23 ns
6-Year Graduation	32.9% (4.01) n = 43	43.9% (4.01) n = 43	F = 3.77 ~ * Eta = .043	33.6% (4.00)	43.2% (4.00)	F = 2.87 ns $F_{(SATM)}$ = 2.41 ns $F_{(SATV)}$ = 2.53 ns

Note. Standard error of the mean in parentheses. Effect size (partial *Eta* squared) is given for significant results.
* p < .05
** p < .01
*** p < .001

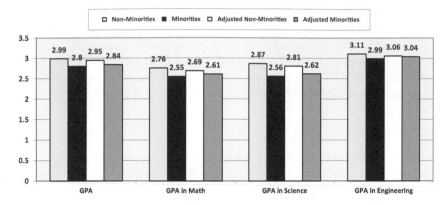

Figure 3.1 Raw and Test-Score Adjusted Academic Performance Components for Minorities and Non-Minorities in NACME Block Grant Institutions

no significant differences in GPA in basic engineering courses. However, minorities exhibited significantly lower overall GPA, GPA in science, and GPA in math. After controlling for test scores, there were no differences in engineering GPA or math GPA, but significant—albeit attenuated—differences in overall GPA and science GPA with minorities achieving lower grades. Controlling for math and verbal SAT scores separately returned similar results.

African American and Non-Minority Students

The patterns observed overall for non-minority and minority grades in engineering remained the same for African Americans compared to non-minorities (Table 3.1). There were no differences in engineering GPA, but African American students achieved lower overall GPA, lower science GPA, and lower math GPA. The test scores of these two groups were significantly different, with African American test scores of 1,106.7 and non-minority test scores of 1,208.6. After controlling for test scores, African American students showed no differences in math GPA or engineering GPA, but still scored lower in overall GPA, and science GPA, although the effect sizes were reduced.

Hispanic and Non-Minority Students

Concerning raw or uncorrected academic performance, there were no differences between Hispanic and non-minority students in core engineering GPA (Table 3.1). However, Hispanic students exhibited lower Overall

GPA, math GPA, and science GPA. There were significant differences in test scores between the two groups with Hispanic students exhibiting lower scores (1,147.8 vs. 1,208.6). After controlling for these differences, the results remained basically the same but of lesser magnitude—no differences in engineering GPA, but significant differences for overall, math and science GPA. There were no significant differences in academic performance between African American and Hispanic students (Table 3.1).

Gender Differences

Table 3.2 displays the results for gender analyses within groups. Although there were no average gender differences in test scores among non-minorities, females exhibited higher overall GPA both before and after corrections for test scores. Among African Americans, there were also no significant gender differences in test scores. Both before and after test score adjustments, African American females exhibited higher overall GPA, math GPA, and engineering GPA. Among Hispanic students where there were also no significant gender differences in test scores, females exhibited higher overall GPA before and after test score adjustments. Grade averages in other course components were somewhat higher among females, but the differences did not reach significance.

Discussion

For underrepresented minorities as a group, their disaggregated grades were lowest in core science courses (of biology and chemistry), even after assuming roughly equal standardized test scores. The differential in science may account for lower overall GPA among minority students. Controlling for math and verbal SAT scores (or equivalents) separately returned similar results. Contrary to findings by Zhang, Anderson, Ohland, and Thorndyke (2004), in no case was SAT-verbal negatively related to a performance measure; in two of three cases where there were differences, SAT-verbal produced stronger rather than weaker results.

When African American and Hispanic students were considered separately and compared with non-minority students of roughly equal test scores, African American grades were significantly lower in science courses but not in math or core engineering courses. Hispanic student grades were lower in all courses except core engineering courses.

By gender, female engineering students of any ethnicity achieved better grades than their male counterparts in one or more course categories, with African American females showing more widespread evidence of better performance. Better performance among female students has been found in several other studies (Borrego et al., 2005; Brawner et al., 2012; Lord et al., 2009; Orr et al., 2014).

Table 3.2 Academic Performance of Minorities and Non-Minorities by Gender in NACME Block Grant Institutions

Measure	Raw Academic Performance			Test Score-Adjusted Academic Performance		
	Non-Minority Males $n = 24$	Non-Minority Females $n = 24$	Significance	Non-Minority Males	Non-Minority Females	Adjusted Significance
SAT	1212.3 (21.0) n = 24	1204.9 (21.0) n = 24	$F = 0.06$ ns	NA	NA	NA
Overall GPA	2.96 (.032) n = 24	3.13 (.032) n = 24	$F = 13.72^{***}$ $Eta = .230$	2.96 (.024)	3.13 (.024)	$F = 27.95^{***}$ $Eta = .383$ $F_{(SATM)} = 32.72^{***}$ $F_{(SATV)} = 18.84^{***}$
Math GPA	2.73 (.066) n = 23	2.87 (.066) n = 23	$F = 2.42$ ns	2.72 (.054)	2.88 (.054)	$F = 4.37^{*}$ $Eta = .092$ $F_{(SATM)} = 5.63^{*}$ $F_{(SATV)} = 2.22$ ns
Science GPA	2.85 (.075) n = 23	2.89 (.075) n = 23	$F = 0.15$ ns	2.85 (.068)	2.89 (.068)	$F = 0.30$ ns $F_{(SATM)} = 0.61$ ns $F_{(SATV)} = 0.02$ ns
Engineering GPA	3.07 (.102) n = 19	3.20 (.102) n = 19	$F = 0.79$ ns	3.06 (.095)	3.21 (.095)	$F = 1.25$ ns $F_{(SATM)} = 1.70$ ns $F_{(SATV)} = 0.51$ ns

(Continued)

Table 3.2 (Continued)

| Measure | Raw Academic Performance | | | Test Score-Adjusted Academic Performance | | |
	Non-Minority Males $n = 24$	Non-Minority Females $n = 24$	Significance	Non-Minority Males	Non-Minority Females	Adjusted Significance
1-Year Retention	80.7% (2.13) $n = 22$	83.9% (2.13) $n = 22$	$F = 1.11$ ns	80.5% (1.91)	84.1% (1.91)	$F = 1.69$ ns $F_{(SATM)} = 2.27$ ns $F_{(SATV)} = 0.64$ ns
4-Year Graduation	22.6% (3.15) $n = 22$	27.8% (3.22) $n = 21$	$F = 1.36$ ns	22.4% (2.54)	27.9% (2.59)	$F = 2.33$ ns $F_{(SATM)} = 3.82\sim$ $Eta = .091$ $F_{(SATV)} = 0.91$ ns
6-Year Graduation	47.7% (3.69) $n = 24$	57.3% (3.69) $n = 24$	$F = 3.39$ ns	47.4% (3.37)	57.6% (3.37)	$F = 4.53^*$ $Eta = .091$ $F_{(SATM)} = 5.28^*$ $F_{(SATV)} = 3.06$ ns

(Continued)

Table 3.2 (Continued)

Measure	Raw Academic Performance			Raw Academic Performance		
	African American Males	African American Females	Significance	African American Males	African American Females	Adjusted Significance
SAT	1103.9 (24.4) n = 23	1109.5 (24.4) n = 23	$F = 0.03$ ns	NA	NA	NA
Overall GPA	2.65 (.060) n = 23	2.85 (.060) n = 23	$F = 5.87^*$ Eta = .118	2.65 (.056)	2.85 (.056)	$F = 6.36^*$ Eta = .129 $F_{(SATM)} = 6.36^*$ $F_{(SATV)} = 6.78^*$
Math GPA	2.37 (.112) n = 22	2.75 (.117) n = 20	$F = 5.45^*$ Eta = .120	2.38 (.110)	2.74 (.115)	$F = 5.24^*$ Eta = .118 $F_{(SATM)} = 5.01^*$ $F_{(SATV)} = 5.33^*$
Science GPA	2.39 (.103) n = 20	2.53 (.105) n = 19	$F = 0.91$ ns	2.39 (.100)	2.52 (.103)	$F = 0.87$ ns $F_{(SATM)} = 0.65$ $F_{(SATV)} = 0.76$
Engineering GPA	2.69 (.157) n = 17	3.14 (.157) n = 17	$F = 4.12\sim$ Eta = .114	2.68 (.153)	3.15 (.153)	$F = 4.66^*$ Eta = .131 $F_{(SATM)} = 4.98^*$ $F_{(SATV)} = 4.31^*$

(Continued)

Table 3.2 (Continued)

| Measure | Raw Academic Performance | | | Test Score-Adjusted Academic Performance | | |
	African American Males	African American Females	Significance	African American Males	African American Females	Adjusted Significance
1-Year Retention	73.4% (4.34) n = 21	80.9% (4.82) n = 17	F = 1.37 ns	73.8% (4.15)	80.4% (4.62)	$F = 1.11$ ns $F_{(SATM)} = 1.15$ ns $F_{(SATV)} = 0.59$ ns
4-Year Graduation	11.9% (3.24) n = 19	12.7% (3.53) n = 16	F = 0.03 ns	11.9% (3.08)	12.7% (3.35)	$F = 0.04$ ns $F_{(SATM)} = 0.04$ ns $F_{(SATV)} = 0.00$ ns
6-Year Graduation	30.1% (5.03) n = 23	36.1% (5.39) n = 20	F = 0.67 ns	30.3% (5.02)	35.9% (5.38)	$F = 0.58$ ns $F_{(SATM)} = 0.78$ ns $F_{(SATV)} = 0.66$ ns

(Continued)

Table 3.2 (Continued)

Measure	Raw Academic Performance			Test Score-Adjusted Academic Performance		
	Hispanic Males	Hispanic Females	Significance	Hispanic Males	Hispanic Females	Adjusted Significance
SAT	1159.8 (23.05) n = 25	1134.7 (24.03) n = 23	$F = 0.57$ ns	NA	NA	NA
Overall GPA	2.77 (.046) n = 25	2.94 (.047) n = 23	$F = 6.73^*$ $Eta = .128$	2.75 (.040)	2.95 (.042)	$F = 11.47^{***}$ $Eta = .203$ $F_{(SATM)} = 10.73^{**}$ $F_{(SATV)} = 10.01^{**}$
Math GPA	2.45 (.096) n = 24	2.64 (.100) n = 22	$F = 1.88$ ns	2.43 (.092)	2.66 (.096)	$F = 2.89$ ns $F_{(SATM)} = 3.63\sim$ $F_{(SATV)} = 2.76$ ns
Science GPA	2.56 (.077) n = 23	2.61 (.079) n = 22	$F = 0.18$ ns	2.54 (.068)	2.63 (.070)	$F = 0.73$ ns $F_{(SATM)} = 1.08$ ns $F_{(SATV)} = 0.57$ ns
Engineering GPA	2.95 (.098) n = 19	3.18 (.104) n = 17	$F = 2.63$ ns	2.94 (.096)	3.19 (.101)	$F = 3.41$ ns $F_{(SATM)} = 3.71\sim$ $F_{(SATV)} = 2.86$ ns

(Continued)

Table 3.2 (Continued)

| Measure | Raw Academic Performance | | | Test Score-Adjusted Academic Performance | | |
	Hispanic Males	Hispanic Females	Significance	Hispanic Males	Hispanic Females	Adjusted Significance
1-Year Retention	80.3% (2.38) n = 23	85.2% (2.49) n = 21	$F = 2.05$ ns	79.9% (2.32)	85.6% (2.43)	$F = 2.75$ ns $F_{(SATM)} = 3.06$ ns $F_{(SATV)} = 2.57$ ns
4-Year Graduation	15.2% (3.19) n = 22	14.6% (3.43) n = 19	$F = 0.02$ ns	14.6% (2.78)	15.3% (2.99)	$F = 0.04$ ns $F_{(SATM)} = 0.07$ ns $F_{(SATV)} = 0.00$ ns
6-Year Graduation	41.7% (5.98) n = 23	46.5% (6.41) n = 20	$F = 0.29$ ns	40.9% (5.99)	47.4% (6.43)	$F = 0.53$ ns $F_{(SATM)} = 0.43$ ns $F_{(SATV)} = 0.56$ ns

Note. Standard error of the mean in parenthesis. Effect size (partial *Eta* squared) is given for significant results.
* $p < .05$
** $p < .01$
*** $p < .001$

Retention to Graduation Rates

Several studies have documented graduation rates for underrepresented minorities, as this is the common yardstick used to judge progress in engineering. Dropout rates hover around 50% for non-minorities and 67% for underrepresented minorities, although significant variation has been documented (e.g., Borrego et al., 2005; Landis, 2005; NACME, 2012; Yoder, 2012). The present study sought to assess retention to graduation rates for participating NACME institutions, that is to assess 1-year retention, 4-year graduation rates, and 6-year graduation rates. The additional objective was to assess these rates before and after adjustments for aggregate institutional standardized test scores. As shown in Table 3.1 and Figure 3.2, for the institutions reporting minority (n = 24) and non-minority (n = 22) 1-year retention rates, the average minority retention rate was 77.1% compared to the non-minority rate of 81.0%, a non-significant difference; 4-year graduation rates were, however, significantly different: 13.6% for minorities vs. 23.3% for non-minorities. Similarly, there were significant differences in 6-year graduation rates—37.3% for minorities, compared to 49.3% for non-minorities. These rates are comparable to those generally reported in the literature, with approximately 67% of minorities failing to graduate in six years, and 50% of non-minorities failing to graduate.

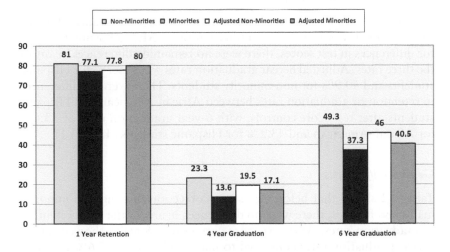

Figure 3.2 Raw and Test-Score Adjusted Retention to Graduation Rates in NACME Block Grant Institutions

Test scores accounted for substantial variance in the retention to graduation rates (see Appendix 3.2). For example, among minorities, test scores accounted for 37.1% of the variance in 1-year retention; 44.5% in 4-year graduation; and 18.1% in 6-year graduation. On average, the variance estimates were even higher among non-minorities—31.7%, 48.9%, and 38.4% respectively. After controlling for differences in test scores, to permit a statement about students of roughly equal ability, there were no significant differences in any of the retention to graduation rates (see Table 3.1 and Figure 3.2). Adjusted 1-year retention rates were 80.0% for minorities and 77.8% for non-minorities; 4-year graduation rates were 17.1% for minorities and 19.5% for non-minorities; and 6-year graduation rates were 40.5% for minorities and 46.0% for non-minorities. Controlling for the math or verbal test scores alone produced similar results.

African American and Non-Minority Students

Comparing African American and non-minority students, for raw retention to graduation rates, there were no differences in 1-year retention, but significant differences in 4-year and 6-year graduation rates. After adjusting these rates for test scores, there were no differences in 1-year retention or 4-year graduation rates. The difference remaining was that African Americans exhibited lower 6-year graduation rates than non-minority students—35.6% vs. 50.1%.

Hispanic and Non-Minority Students

Comparing Hispanic and non-minority students, Hispanic students showed only lower raw 4-year graduation rates. After adjusting the rates for differences in test scores, there were no remaining differences in any of the three rates. Adjusted 6-year graduation rates were 45.7% for Hispanic students and 50.9% for non-minority students. There were no differences in retention to graduation rates between African American and Hispanic students after test-score controls, with 6-year graduation rates of 33.6% for African American and 43.2% for Hispanic students (Table 3.1).

Gender Differences

Table 3.2 shows that before test score corrections among non-minority males and females, there were no differences in any of the three retention to graduation rates. After test score corrections, females showed higher 6-year graduation rates compared to males (47.5% vs. 57.6%). Among African Americans, there were no significant gender differences in retention to graduation rates before or after test score corrections. Female rates were somewhat higher, but the differences did not reach statistical significance. Similarly, among Hispanic students, there were no gender differences in retention to graduation rates before or after test score corrections. Female rates were generally higher, but not significantly so.

Discussion

To summarize, the analysis of raw rates showed that there were no differences in 1-year retention rates, but statistically significant differences in favor of non-minority students in 4- and 6-year graduation rates. However, non-minority standardized test scores were significantly higher than those of minority students, thus obscuring real differences in matched abilities. When a method of adjusting for, or controlling for, differences in test scores was employed, there were no differences at all between the two comparison groups in any of the three retention to graduation rates. Thus, when considering students of roughly equal ability, there were no significant differences in retention to graduation rates. By ethnicity, African American students still showed lower 6-year graduation rates after test score controls. By gender, non-minority females achieved significantly higher 6-year graduation rates than non-minority males. However, no such gender advantages were observed among African American or Hispanic students, despite the previously observed better grades among these female students.

Ranking NACME Engineering Institutions

In the same way that we consider raw and adjusted retention to graduation rates for minorities and non-minorities, we are also able to rank engineering institutions according to the highest rates for minorities and non-minorities using raw rates as well as rates adjusted for differences in test scores. Schools with the best students typically have the highest graduation rates, but the concern here was which schools do the most with what they have. Or, in other words, which schools have the best programs rather than just the best students, as judged by test scores? Examining adjusted scores permits just a statement.

Table 3.3 shows the ranking of the top institutions based on the raw (i.e., unadjusted) retention to graduation rates for minority students. Using 6-year rates was the most reliable measure and the one reported by the greatest number of institutions. In this case, the top five institutions were: Georgia Tech; Virginia Tech; Missouri Tech; University of Colorado, Boulder; and University of Michigan. Table 3.3 also shows that after an adjustment for average minority student test scores for each institution, the top five institutions became: Georgia Tech; University of Texas at El Paso' North Carolina A&T; Kettering Institute; and Virginia Tech. Table 3.3 displays the top ten schools according to their graduation rates for minority students, along with the minority and non-minority rates by institution, as well as the differences in minority/non-minority rates.

Table 3.3 Top Ten Institutions for Raw and Test-Score-Adjusted 6-Year Graduation Rates, Ranked by Minority Rates

Rank	School	Raw Minority Rate	Raw Non-Minority Rate	Discrepancy in Rate	Parity Rank
	Raw 6-Year Graduation Rates				
1	Georgia Tech	78.2%	82.7%	-4.5	4
2	Virginia Polytechnic Institute	60.0%	67.7%	-7.7	5
3	Missouri Technological Institute	56.0%	58.0%	-2	3
4	University of Colorado, Boulder	51.0%	60.0%	-9	6
5	University of Michigan	51.0%	79.0%	-28	9
6	University of Arkansas	50.0%	43.3%	+6.7	2
7	Kettering Institute	48.1%	68.6%	-20.5	8
8	North Carolina A&T	48.0%	31.0%	+17	1
9	New Jersey Institute of Technology	48.0%	57.0%	-9	6
10	University of Texas, El Paso	41.0%	51.9%	-10.9	7

(Continued)

Table 3.3 (Continued)

Rank	School	Adjusted Minority Rate	Adjusted Non-Minority Rate	Discrepancy in Rate	Parity Rank
		Test-Score Adjusted 6-Year Graduation Rates			
1	Georgia Institute of Technology	67.1%	62.9%	+4.2	4
2	University of Texas, El Paso	59.1%	62.5%	-3.4	8
3	North Carolina A&T	58.4%	39.7%	+18.7	1
4	Kettering Institute	58.0%	70.6%	-12.6	9
5	Virginia Polytechnic Institute	56.4%	58.7%	-2.3	6
6	Missouri Technological Institute	56.4%	52.1%	+4.3	3
7	University of Arkansas	54.1%	38.9%	+15.2	2
8	New Jersey Institute of Technology	53.4%	55.8%	-2.4	7
9	California State University, Los Angeles	50.1%	46.7%	+3.4	5
10	Prairie View A&M University	48.8%	NA	NA	

In Search of Parity in 6-Year Graduation Rates

One of NACME's concerns is to facilitate parity in engineering graduation rates. Thus, one consideration for consortium membership is equal or near-equal graduation rates for minority and non-minority students (see Table 3.3). Using raw 6-year graduation rates for the 24 institutions with rates for both minorities and non-minorities, the average difference in graduation rates was −11.6% (i.e., minority rates were 11.6% *lower* than those for non-minority students), with a range of −35.0%−+17.0%. When institutions in the top ten of minority graduation rates were considered, minorities averaged 6.79% *lower* rates than non-minorities, with a range of −28.0%−+17.0%.

Parity rates were negatively related to the average non-minority test score for the University at a marginal level ($r = -.349$), meaning the higher the average student test score, the more disparate the graduation rates. Parity rates were also negatively related to the average minority test score, but at a non-significant level. Using 6-year graduation rates adjusted for differences in SAT or equivalent scores, minorities averaged rates that were 5.26% *lower* than non-minorities, with a range of −21.6%−+18.8%. Considering those institution with the ten highest minority graduation rates, minorities averaged rates of 1.39% *higher* than non-minorities. In short, with 6-year graduation rates adjusted for differences in test scores, average graduation rates for minorities and non-minorities moved closer to parity. Further, for the schools with the best test score-adjusted minority graduation rates, near-equal parity was achieved, at least on the average, for the top ten schools.

Discussion

According to these procedures, the institutions with the best adjusted 6-year graduation rates were: Georgia Tech; University of Texas, El Paso; North Carolina A&T; Kettering Institute; and Virginia Tech. Why do these institutions top the list? We can speculate. Georgia Tech's committed leadership has been extolled for many years (Maton, Watkins-Lewis, Beason, & Hrabowski, 2015; Ransom, 2015; Sidbury, Johnson, & Burton, 2015), and its dual degree programs with several HBCUs are alleged to provide it with a hidden Black college advantage (Malcolm-Piquex & Malcolm, 2015). UTEP is a Minority-Serving Institution, and although Hispanic-Serving Institutions are not known for conferring the same educational advantage as HBCUs because their designation is statistical rather than mission-based (Bridges et al., 2008; Contreras, Malcolm, & Bensimon, 2008; Nunez, Hurtado, & Galdeano, 2015), UTEP's excellence in first year programming and STEM instruction has been widely acknowledged, and UTEP was designated a national Model Institution of Excellence by the National Science Foundation. North Carolina A&T has been among the top producers of African American engineers for many years (Ransom, 2015), as have a number of other HBCUs. While these schools have been accused of low retention rates and remaining top producers due to larger absolute numbers of enrolled students (Reichert & Absher, 1997), this analysis points to an effective retention of students in engineering at N.C. A&T. Kettering Institute has a unique curriculum with alternating semesters of coursework and industry internships. Such hands-on immersion in the work of engineering has been cited as a major success factor for engineering student success (Fifolt & Abbott, 2008; Fleming, 2016). Virginia Tech has been extolled and sometimes berated for its deliberate efforts to pave the way for African Americans to attend this formerly all-White institution (Crichton, 2003), for its presidential support for doubling the enrollment of its underrepresented minorities by partnering with 15 Virginia high schools (Korth, 2016), and for supporting the development of effective retention programs staffed by administrative leaders with noted scholarship records (e.g., Amelink & Meszaros, 2011; Lee, Brozina, Amelink, & Jones, 2017; Lent et al., 2015, 2016; Van Aken, Watford, & Medina-Borja, 1997; Watford, 2011).

Conclusion

NACME participating institutions were invited to become block grant members because they have demonstrated an ability to recruit, support, and retain minorities in engineering. They have Minorities in Engineering support programs, have minority retention to graduation rates that are above the national average, and/or have demonstrated an ability to graduate minority engineering students at near-equal rates to their non-minority students. Twenty-six of these institutions responded to a request for aggregate institutional data on their minority and non-minority students for a series of statistical comparisons to assist NACME in following the progress of its member institutions. This analysis focused not only on raw grade averages and retention to graduation rates, but also on these average rates adjusted for group differences in test scores. In this way, the institutional outcomes make a statement about groups of students with roughly equal abilities as measured by math and verbal test scores.

The institutional analyses showed that with aggregate test scores adjusted, there were fewer differences between minority and non-minority students in retention to graduation rates compared to other reports that did not control for differences in test scores. The raw, unadjusted rates were similar to those frequently reported in the literature, with approximately 67% of minority students failing to graduate in six years, and roughly 50% of non-minority students failing to graduate during this same period. However, 6-year graduation rates adjusted for student test scores showed that 6-year graduation rates were 40.5% for minorities and 46.0% for non-minorities—a non-significant difference, indicating that 59.5% of minorities and 54.0% of non-minorities fail to graduate. This also suggests that analyses of minority and non-minority students with near-equal test scores in the same institutions would minimize underperformance outcomes among minority students and provide a more accurate profile of ethnic differences.

While this is the picture for underrepresented minority students as a group, there were differences to note when they were disaggregated by ethnic group. Specifically, African American students showed significantly lower 6-year graduation rates compared to non-minority students, even after test-score adjustments, of 35.6%–50.1%. Other reports have documented graduation rates that were lowest among this group (NACME, 2012; Yoder, 2012).

Previous studies have reported lower grade averages among underrepresented minority students in engineering. The present study found similar results. Indeed, there was stronger evidence of underperformance among minorities in grades, with or without controls for test scores, than was the case for retention to graduation rates. Minority students did not perform lower in core engineering courses, and this finding cut across ethnicity. However, minority students performed lower principally in science courses as well as in overall grade averages. African American students did not perform lower in math courses, although Hispanic students did—after test score adjustments. The strongest underperformance in science courses raises questions as to why this is so. While the systemic lack of exposure to and/or underperformance in science experienced by these groups of students suggests itself as an explanation (Dalton, Ingels, Downing, Bozick, & Owings, 2007; Lutkus, Lauko, & Brockway, 2006), the present study provides no further clues.

Continued investigation of gender differences in engineering education seem warranted. Most of the studies reviewed found that women in undergraduate engineering achieve better grades and have higher persistence rates than their male counterparts, a finding that appears to transcend ethnicity. This study concurs in finding that women of any ethnicity get better grades, with the strongest and most widespread advantage going to African American women. Yet, higher graduation rates were found only among non-minority women compared to men. Both Black and Hispanic female graduation rates were slightly higher than their male counterparts, but not significantly so. Why a minority female GPA advantage does not yield higher graduation rates as it does for non-minority women raises an interesting, but here unanswered, question.

Previous reports of effective engineering institutions for minority students consistently cite HBCUs as overrepresented among the top producers of minority engineering bachelor's degrees. The present highlighting of the top five institutions with the highest minority graduation rates, with test scores held constant, includes an HBCU, North Carolina A&T, but also indicates that a variety of institutional types and characteristics can be effective for minority students—the racial and ethnic comfort or support provided by Minority-Serving Institutions (both Black and Hispanic), a hidden Black college advantage provided by dual degree programs, the power of the hands-on engineering experience provided by a cooperative curriculum, forceful and visible leadership on behalf of underrepresented minority students and/or support programs, and the importance of administrator-scholars. In other words, there appear to be multiple pathways available for engineering institutions seeking to be leaders in engineering retention. Furthermore, the analysis of parity rates that converge in the top institutions indicates that engineering schools effective in retaining underrepresented minority students are also effective in retaining all students.

The findings also raise the question of whether this study provides any other evidence of adjustment disadvantages for minority students in NACME engineering schools in general—adjustment disadvantages of the kind so often discussed in the general literature on minority students in college (Ferguson, 2003; Fries-Britt & Turner, 2002; Oseguera, Hurtado, Denson, Cerna, & Saenz, 2006; Pascarella & Terenzini, 2005; Rovai, Gallien, & Wighting, 2005; Smith, 2009; Solórzano, Ceja, & Yosso, 2000). The underperformance of minorities in college is taken as evidence of poor adjustment, racial isolation, or discrimination against African American and Hispanic students (e.g., Johnson et al., 2007; Karkouti, 2016; Lewis, Mendenhall, Harwood, & Hunt, 2012; Tannenbaum & Ruck, 2007). Several studies have documented the negative effects on the performance of minority engineering students due to the perception of racism (Brown, Morning, & Watkins, 2005; Lee, Matusovich, & Brown, 2014; Yoder, 2012). However, underperformance after test scores are held constant provides more compelling evidence of environmental duress, as has been shown in a long series of such studies (Steele, 1997; Steele & Aronson, 1995; Watson, 1972). In this study, African American students underperform on two of four measures of GPA and 6-year graduation rates; Hispanic students underperform on three measures of GPA—after test scores were held constant. Furthermore, both groups of minority women achieved better grades than their male counterparts, but these higher grades did not translate into higher graduation rates (as was true for non-minority women). Such patterns could be indications of the differential treatment or support issues described in other studies.

A final way in which adjustment issues could be indicated is in differential patterns of test score to performance outcome correlations. The patterns observed showed that the strength of the average correlations with performance measures was attenuated among minority students, such that average variance accounted for among minorities was 26.8% compared to 36.6% for non-minorities. When broken down into minority ethnic groups average variance accounted for was lower still for

African American and Hispanic students—6.4% and 11.8%, respectively. In previous studies, weaker correlations of this kind have been found generally for Black students in White colleges, but not in HBCUs (Fleming, 2002, 2004), with the conclusion that environmental issues, such as adjustment, interfere with the normal translation of test score potential into academic outcomes. Such possible issues bear further investigation.

A limitation of this set of analyses is that they are based on aggregate institutional data, when student-specific data would provide the most accurate, if less feasible, account. Also, ethnicity data was reported separately by gender, such that combining the two genders to create ethnicity totals may provide less accurate data than an institutional total. Nevertheless, this attempt has provided data from the greatest number of NACME institutions so far, and periodic assessments of this kind would constitute an important step in monitoring the academic outcomes for minorities in these engineering institutions.

References

Adelman, C. (1998). *Women and men of the engineering path: A model for analyses of undergraduate careers*. Washington, DC:U.S. Department of Education.

Amelink, M., & Meszaros, P. (2011). A comparison of educational factors promoting or discouraging the intent to remain in engineering by gender. *European Journal of Engineering Education, 36*(1), 47–62.

Araque, F., Roldan, C., & Salguero, A. (2009). Factors influencing university dropout rates. *Computers and Education, 53*(3), 563–574.

Atwaters, S, Y., Leonard, J. D., & Pearson, W. (2015). Beyond the black-white minority experience: Undergraduate engineering trends among African Americans. In J. B. Slaughter, Y. Tao, & W. Pearson, Jr. (Eds.), *Changing the face of engineering: The African American experience* (pp. 149–188). Baltimore, MD: Johns Hopkins University Press.

Borrego, M. J., Padilla, M. A., Zhang, G., Ohland, M. W., & Anderson, T. J. (2005). *Graduation rates, grade-point average, and changes of major of female and minority students entering engineering*. Proceedings, ASEE/IEEE Frontiers in Education Conference, October 19–22, Indianapolis, IN.

Brawner, C. E., Camacho, M. M., Lord, S. M., Long, R. A., & Ohland, M. W. (2012). Women in industrial engineering: Stereotypes, persistence, and perspectives. *Journal of Engineering Education, 101*(2), 288–318.

Bridges, B. K., Kinzie, J., Laird, T. F. N., & Kuh, G. D. (2008). Student engagement and student success at historically Black and Hispanic-serving institutions. In M. Gasman, B. Baez, & C. S. V. Turner (Eds.), *Understanding minority-serving institutions* (pp. 217–236). Albany, NY: State University of New York Press.

Brown, A. R., Morning, C., & Watkins, C. (2005). Influence of African American engineering student perceptions of campus climate on graduation rates. *Journal of Engineering Education, 94*(4), 263–271.

Camacho, M. M., & Lord, S. M. (2013). *The borderlands of education: Latinas in engineering*. Lanhan, MD: Lexington Books.

Chen, X., & Weko, T. (2009). *Students who study science, technology, engineering and math (STEM) in postsecondary education.* Washington, DC: U.S. Department of Education.

Chubin, D. E., Donaldson, K., Olds, B., & Fleming, L. (2008). Educating generation net: Can U.S. engineering woo and win the competition for talent? *Journal of Engineering Education, 97*(3), 245–258.

Chubin, D. E., May, G. S., & Babco, E. L. (2005). Diversifying the engineering workforce. *Journal of Engineering Education, 94*(1), 73–86.

Cole, D. (2008). Constructive criticism: The role of student-faculty interactions on African American and Hispanic students' educational gains. *Journal of College Student Development, 49*(6), 587–605. doi: 10.1353/csd.0.0040.

Constantine, M. G., Wilton, L., & Caldwell, L. D. (2003). The role of social support in moderating the relationship between psychological distress and willingness to seek psychological help among Black and Latino college students. *Journal of College Counseling, 6*(2), 155–165.

Contreras, F. E., Malcolm, L. E., & Bensimon, E. M. (2008). Hispanic-serving institutions: Closeted identity and the production of equitable outcomes for Latino/a students. In M. Gasman, B. Baez, & C. S. V. Turner (Eds.), *Understanding minority-serving institutions* (pp. 71–90). Albany, NY: State University of New York Press.

Crichton, J. (2003). Paving the way: African Americans at Virginia Tech. *Virginia Tech Magazine,* Spring. Retrieved from www.vtmag.vt.edu/spring03/feature2.html.

Dalton, B., Ingels, S. J., Downing, J., Bozick, R., & Owings, J. (2007). *Advanced mathematics and science coursetaking in the spring high school senior classes of 1982, 1992 and 2004: Statistical analysis report.* Washington, DC: U.S. Department of Education.

Ferguson, R. F. (2003). Teachers' perceptions and expectations and the black-white test score gap. *Urban Education, 38*(4), 460–507.

Fifolt, M. M., & Abbott, G. (2008). Differential experiences of women and minority engineering students in a cooperative education program. *Journal of Women and Minorities in Science and Engineering, 14*(3), 253–267.

Fleming, J. (2002). Identity and achievement: Black ideology and the SAT in African American college students. In W. R. Allen, M. B. Spencer, & C. O'Connor (Eds.), *African American education: Race community, inequality and achievement* (pp. 77–92). Burlington, MA: Elsevier Science.

Fleming, J. (2004). The significance of historically black colleges for high achievers: Correlates of standardized test scores in African American students. In M. C. Brown II & K. Freeman (Eds.), *Black colleges: New perspectives on policy and practice* (pp. 29–52). Westport, CT: Praeger.

Fleming, J. (2016). *Success factors for minorities in engineering: Analysis of focus group mini-surveys.* Proceedings of the 2016 Annual Conference of the American Society of Engineering Education (ASEE), New Orleans, LA, June 26–29. Retrieved from https://peer.asee.org/16161.

Fries-Britt, S., & Turner, B. (2002). Uneven stories: The experiences of successful Black collegians at a historically Black and a traditionally White campus. *Review of Higher Education, 25*(3), 315–330.

Gasman, M. (2008). Minority-serving institutions: A historical backdrop. In M. Gasman, B. Baez, & C. S. V. Turner (Eds.), *Understanding minority-serving institutions* (pp. 18–27). Albany, NY: State University of New York Press.

Ishitani, T. T. (2006). Studying attrition and degree completion behavior among first generation students in the United States. *Journal of Higher Education*, 77(5), 861–885.

Johnson, D. R., Soldner, M., Leonard, J. B., Alvarez, P., Inkelas, K. K., Rowan-Kenyon, H. T., & Longerbeam, S. D. (2007). Examining sense of belonging among first-year undergraduates from different racial/ethnic groups. *Journal of College Student Development*, 48(5), 525–542.

Karkouti, I. M. (2016). Black students' educational experiences in predominantly white institutions: A review of related literature. *The College Student Journal*, 50(1), 59–70.

Korth, R. (2016). Va Tech president calls on university to double minority enrollments. *The Roanoke Times*, November 7.

Landis, R. (2005). *Retention by design*. New York: National Action Council for Minorities in Engineering.

Lee, W. C., Brozina, C., Amelink, C. T., & Jones, B. D. (2017). Motivating incoming engineering students with diverse backgrounds: Assessing a summer bridge program's impact on academic motivation. *Journal of Women and Minorities in Science and Engineering*, 23(2), 121–145.

Lee, W. C., Matusovich, H. M., & Brown, P. R. (2014). Measuring underrepresented student perceptions of inclusions within engineering departments and universities. *International Journal of Engineering Education*, 30(1), 150–165.

Lee, J. M., & Rawls, A. (2010). *The college completion agenda: 2010 progress report*. New York: The College Board. Retrieved from http://completionagenda. college board.org/sites/default/files/reports_pdf.Progress_Report_2010.pdf.

Lent, R. W., Miller, M. J., Smith, P. E., Watford, B. A., Hui, K., & Lim, R. H. (2015). Social cognitive model of adjustment to engineering majors: Longitudinal test across gender and race/ethnicity. *Journal of Vocational Behavior*, 86(1), 77–85.

Lent, R. W., Miller, M. J., Smith, P. E., Watford, B. A., Lim, R. H., & Hui, K., (2016). Social cognitive predictors of academic persistence and performance in engineering: Applicability across gender and race/ethnicity. *Journal of Vocational Behavior*, 94(1), 79–88.

Lewis, J. A., Mendenhall, R., Harwood, S. A., & Hunt, M. B. (2012). Coping with gendered racial microaggressions among Black women college students. *Journal of African American Studies*, 17(1), 51–73.

Lord, S. M., Camacho, M. M., Layton, R. A., Long, R. A., Ohland, M. W., & Wasburn, M. H. (2009). Who's persisting in engineering? A comparative analysis of female and male Asian, black, Hispanic, Native American, and white students. *Journal of Women and Minorities in Science and Engineering*, 15(2), 167–190.

Lutkus, A. D., Lauko, M., & Brockway, D. (2006). *The nation's report card: Trial urban district assessment of grades 4 and 8: Science 2005*. Washington, DC: U.S. Department of Education, National Assessment of Educational Progress.

Malcolm-Piquex, L. E., & Malcolm, S. (2015). African American women and men into engineering: Are some pathways smoother than others? In J. B. Slaughter, Y. Tao, & W. Pearson, Jr. (Eds.), *Changing the face of engineering: The African American experience* (pp. 90–119). Baltimore, MD: Johns Hopkins University Press.

Marra, R. M., Rodgers, K. A., & Shen, D. (2012). Leaving engineering: A multi-year single institution study. *Journal of Engineering Education*, 101(1), 6–27.

Maton, K. I., Watkins-Lewis, K. M., Beason, T., & Hrabowski, F. A. III. (2015). Enhancing the number of African Americans pursuing the Ph.D. in engineering: Outcomes and processes in the Meyerhoff Scholarship Program. In J. B. Slaughter, Y. Tao, & W. Pearson, Jr. (Eds.), *Changing the face of engineering: The African American experience* (pp. 354–386). Baltimore, MD: Johns Hopkins University Press.

NACME (National Action Council for Minorities in Engineering). (2012). African Americans in engineering. *Research & Policy Brief*, 2(4, September), 1–2.

Nunez, A. M., Hurtado, S., & Galdeano, E. C. (2015). Why study Hispanic-serving institutions? In A. M. Nunez, S. Hurtado, & E. C. Galdeano (Eds.), *Hispanic-serving institutions: Advancing research and transformative practice* (pp. 1–22). New York: Routledge.

Ohland, M. W., Brawner, C. E., Camacho, M. M., Layton, R. A., Long, R. A., Lord, S. M., & Wasburn, M. H. (2011). Race, gender, and measures of success in engineering education. *Journal of Engineering Education*, 100(2), 225–252.

Orr, M. K., Lord, S. M., Layton, R. A., & Ohland, M. W. (2014). Student demographics and outcomes in mechanical engineering in the US. *International Journal of Mechanical Engineering Education*, 42(1), 48–60.

Oseguera, L., Hurtado, S., Denson, N., Cerna, O., & Saenz, V. (2006). The characteristics and experiences of minority freshmen committed to biomedical and behavioral science research careers. *Journal of Women and Minorities in Science and Engineering*, 155–177.

Pascarella, E. T., & Terenzini, P. T. (2005). *How college affects students*. San Francisco: Jossey-Bass.

Ransom, T. (2015). Clarifying the contributions of historically black colleges and universities in engineering education. In J. B. Slaughter, Y. Tao, & W. Pearson, Jr. (Eds.), *Changing the face of engineering: The African American experience* (pp. 120–148). Baltimore, MD: Johns Hopkins University Press.

Reichert, M., & Absher, M. (1997). Taking another look at educating African American engineers: The importance of undergraduate retention. *Journal of Engineering Education*, 86(3), 241–253. doi: 10.1002/j.2168–9830.1997.tb00291.x.

Rovai, A. P., Gallien, L. B., & Wighting, M. J. (2005). Cultural and interpersonal factors affecting African American academic performance in higher education. *Journal of Negro Education*, 74(4), 359–370.

Sidbury, C. K., Johnson, J. S., & Burton, R. Q. (2015). Spelman's dual-degree engineering program: A path for engineering diversification. In J. B. Slaughter, Y. Tao, & W. Pearson, Jr. (Eds.), *Changing the face of engineering: The African American experience* (pp. 335–353). Baltimore, MD: Johns Hopkins University Press.

Slaton, A. (2011). *Race, rigor, and selectivity in U.S. engineering: The history of an occupational color line*. Cambridge, MA: Harvard University Press.

Slaughter, J. B. (2015). Introduction. In J. B. Slaughter, Y. Tao, & W. Pearson, Jr. (Eds.), *Changing the face of engineering: The African American experience* (pp. 1–9). Baltimore, MD: Johns Hopkins University Press.

Smith, C. P. (1992). *Motivation and personality: Handbook of thematic content analysis*. New York: Cambridge University Press.

Smith, W. A. (2009). Campuswide climate: Implications for African American students. In L. C. Tillman (Ed.), *The SAGE handbook of African American education* (pp. 297–309). Thousand Oaks, CA: Sage.

Solórzano, D., Ceja, M., & Yosso, T. (2000). Critical race theory, racial microaggressions, and campus racial climate: The experiences of African American college students. *Journal of Negro Education, 69*(1/2), 60–73.

Steele, C. M. (1997). A threat in the air: How stereotypes shape intellectual identity. *American Psychologist, 52*(6), 613–629.

Steele, C. M., & Aronson, J. (1995). Stereotype threat and the intellectual test performance of African Americans. *Journal of Personality and Social Psychology, 69*(5), 797–811.

Suresh, R. (2006/2007). The relationship between barrier courses and persistence in engineering. *Journal of College Student Retention, 8*(2), 215–239.

Swail, W. S., & Chubin, D. E. (2007). *An evaluation of the NACME block grant program.* Virginia Beach, VA: Educational Policy Institute.

Tennenbaum, H. R., & Ruck, M. D. (2007). Are teacher's expectations different for racial minority than for European American students? A meta-analysis. *Journal of Educational Psychology, 99*(2), 253–273.

Van Aken, E. M., Watford, B., & Medina-Borja, A. (1999). The use of focus groups for minority engineering program assessment. *Journal of Engineering Education, 88*(3), 333–343.

Watford, B. (2011, Summer). This time, let's make progress: ASEE has a new chance to promote diversity. *Prism.* Retrieved from www.prism-magazine.org/summer11/last_word.cfm.

Watson, P. (1972). IQ—The racial gap. *Psychology Today,* September, 48.

Yoder, B. (2012). *Going the distance in engineering education: Best practices and strategies for retaining engineering, engineering technology, and computing students.* Washington, DC: American Association for Engineering Education (ASEE).

Zhang, G., Anderson, T. J., Ohland, M. W., & Thorndyke, B. R. (2004). Identifying factors influencing engineering student graduation: A longitudinal and cross-institutional study. *Journal of Engineering Education, 93*(4), 313–320.

Appendix 3.1
Methodological Details

The following measures were requested:

1. The number of undergraduate students enrolled, broken down by year.
2. A breakdown of the total enrollment by ethnicity; i.e., the number of African American, Latino/a, and Native American students.
3. A breakdown of the total number by gender within ethnic group.
4. Average SAT (math and verbal) or composite ACT scores for enrolled students.
5. Average GPA for the 2012 cohort of students, with subcomponents for math GPA, science GPA, and GPA in core engineering courses.
6. Average retention to graduation statistics; i.e., 1-year retention, 4-year graduation and 6-year graduation rates.

All institutions were not able to report data for all measures, with the lowest numbers reporting GPA in core engineering courses. Numbers of institutions reporting data for non-minorities ranged from 19 to 24, 19–24 each for non-minority males and females; 20–26 for minorities as a whole; 17–23 for African American males and females and 34–46 for the African American group; 19–25 for Hispanic males, 17–23 for Hispanic females, and 36–48 for Hispanics as a group.

Data Analysis

To answer the research questions, two types of analysis were conducted. First, descriptive statistics permit the reporting of average outcomes.

Second, analysis of variance was used to determine comparative academic performances of non-minorities and minorities, both overall, by ethnic group, and by gender within ethnic group. Note that because ethnic data was requested only by gender, combining the gender groups doubled the number of institutions.

Third, analysis of covariance was used to determine comparative differences in outcomes after any differences in standardized test scores were controlled.

Fourth, for minority and non-minority students, retention and graduation rates were adjusted for differences in test scores. For example, adjusted 6-year graduation rates, given test scores, were calculated on the basis of the overall regression of graduation rates on test scores. For each school, the graduation rate was multiplied by a constant: the overall correlation coefficient of graduation rates with test scores, times the ratio of the overall standard deviations of graduation rates and test scores (SD of graduation rates divided by SD of test scores). Each school's adjusted graduation rates were then subtracted from their actual scores to give corrected rates that were uncorrelated with test scores. An added constant returns the rates to positive numbers. Graduation rates were adjusted for differences in test scores by means of the following computational formula (see Smith, 1992, p. 534):

$$Y' = Y - \left[\left(r_{xy} * SD_y / SD_x * X \right) \right]$$

Where:

Y' = the corrected outcome score (such as 6-year retention) for a given individual/school

Y = the original, uncorrected outcome score (such as 6-year retention) for a given individual/school

r_{xy} = the correlation between the outcome measure and test scores for the whole group

SD_y = the standard deviation of the uncorrected outcome score for the whole group

SD_x = the standard deviation of test scores for the whole group

* = multiply by

/ = divided by

X = test scores for a given individual/school

Note that the magnitude of the difference in subsequent corrected values depends on the strength of the correlation between the two variables at issue—in this case, retention to graduation rates and test scores. See Appendix 3.2 for the test score correlations with all performance measures in the investigation and average variance accounted for by them. Following the correction, there will be a zero-order correlation between the variables. Note that the corrected values typically return negative numbers; as such, a constant is required to bring the values into the positive range. The constant chosen involves a judgment based on the lower to upper range of the original distribution. In this case, the constant added was sufficient to provide results as close as possible to those provided by analysis of covariance. Also, note that the exact values of a corrected distribution differ by the target group because of differences in the test score to performance criterion correlation. Thus, a distribution corrected among students of one ethnic group will differ from one that combines minority and non-minority students.

Fifth, the rankings in retention to graduation rates obtained from the previous analysis permit the determination of institutions with the highest retention to graduation rates.

Appendix 3.2

Patterns of Correlations Between Test Scores and Performance Outcomes in NACME Block Grant Institutions

Group	1-Year Retention Rate	4-Year Graduation Rate	6-Year Graduation Rate	Overall GPA	Math GPA	Science GPA	Engineering GPA	Average Correlation/ Variance Accounted for
Minorities	.609** n = 24	.667*** n = 24	.425* n = 25	.476* n = 26	.548** n = 24	.516** n = 24	.387 ns n = 20	.518 R^2 = 26.8%
Non-Minorities	.563** n = 22	.699*** n = 22	.620*** n = 24	.779*** n = 24	.614* n = 23	.532* n = 23	.453* n = 19	.609 R^2 = 37.1%
Minorities & Non-Minorities Grouped	.597*** n = 46	.723*** n = 46	.576*** n = 49	.674*** n = 50	.625*** n = 47	.594*** n = 47	.446* n = 39	.605 R^2 = 36.6%
African Americans (Male and Female Grouped)	.344* n = 38	.357* n = 35	.175 ns n = 43	.363* n = 46	.255 ns n = 42	.269 ns n = 39	.256 ns n = 34	.252 R^2 = 6.4%
Hispanics (Male & Female Grouped)	.254 ns n = 44	.514*** n = 41	.164 ns n = 43	.429* n = 48	.291* n = 46	.488*** n = 45	.259 ns n = 36	.343 R^2 = 11.8%

* $p < .05$
** $p < .01$
*** $p < .001$

4 Profile of Minority Engineering Students

Analysis of Focus Group Conversations

Summary

From 11 universities, 176 students participated in 45-minute focus group sessions that explored their engineering experiences from first budding interest to career plans. The students were not selected randomly, but invited because they were involved in program activities and known to their faculty and staff. They are successful to the extent that they were engaged in their studies and program activities. They were an inspiring group of students who gravitated to engineering largely by inclination or family influence, who were all able to do math and to whom computer applications are second nature, who were often groomed by exposure to STEM programs in secondary school and summer bridge, who work in groups and therefore have access to considerable help when they need it, who thrive in student professional organizations, and who pay little attention to racism, failure, or frequent setbacks. They are problem solvers and doers who want to make things, build things to have an impact on the world, or in some cases to engage in research with the same effect. This is a vanguard of minority students poised to improve their own worlds and the world around them.

Introduction

What we know of engineering personalities and types comes from research describing engineering student learning styles. Much of this work is concerned with broadening the traditional lecture teaching methods in engineering to include more active, hands-on methods that appeal to a wider range of students. These studies describe engineering students as comfortable with mathematics, introverted, intelligent, creative, effective, and good organizers, or as single-minded, brusque, and not people oriented. (Broberg, Griggs, & I-Hai Lin, 2006; Felder & Silverman, 1988; Godfrey & Parker, 2010; Scissons, 1979). Harrison, Tomblen, and Jackson, (2006) dispute the introverted stereotype and suggest they are simply impersonal. Nonetheless, introverted types are said to function well in the traditional teaching modes, but hands-on instruction—including project or problem-based instruction, hyper media, and web access— offer effective teaching modes for all learning styles (Zywno & Waalen, 2002). Virtually nothing is known about the personality orientation or learning styles of minority engineering students, although broadening teaching styles has been shown to improve minority student performance as much or more than other students in the general sciences. The focus group approach seemed suited for gathering a general profile of minority engineering students.

As a prelude to the student level analyses to follow, a series of focus groups were conducted with minority engineering students. In order to design an informative survey instrument, greater familiarity with these students was required. To that end, students from 11 universities were chosen for separate group interviews. The goal was to have students comment on their entry into engineering studies and educational experiences from their first interest in the subject to their vision of an engineering career. A sample of eight universities was chosen because of having relatively high or low retention and graduation rates established in Chapter 2. While the hope was to choose in this way to help select programmatic success factors, in truth, some of the lower rates were due to artifacts of two-tiered entry into the engineering school or to the absence of 4-year graduation rates in schools with 5-year programs. Also, three additional universities in Texas were chosen because they provided the comparison of a Predominantly White Institution, an Historically Black College or University, and a Hispanic-Serving Institution. While it is not feasible to apply statistical analysis to focus group discussions, a mini-questionnaire was also administered for this purpose. This sample, although not a randomly selected one, does provide an acquaintance with students who are more engaged in their engineering program activities.

Focus groups were chosen as the initial strategy for developing solutions to the major issues. The focus group process of collecting data, initially devised in the social sciences, has achieved it greatest popularity in marketing research (Heiskanen, Jarvela, Pulliainen, Saastamoinen, & Timonen, 2008), but has re-emerged as a new tool in student evaluation research (Krueger, 1994; Morgan, 1993; Stage, 1992; Stewart, Shandasani, & Rook, 2006). Focus groups have the advantage of allowing individuals greater comfort, reducing the anxiety aroused in one-on-one interviews, and permitting them to generate a wide range of responses by engaging in conversation with each other. The disadvantages of conversations potentially dominated by strong personalities and the inability to conduct correlational analyses are offset by the need of exploratory stages of evaluation to search for an approach to the major issues. Byers and Wilcox (1988) suggested that focus groups may be the best way of obtaining the data necessary for exploratory research when little is known beforehand. Boddy (2005) made a distinction between focus groups and group discussions. Focus groups, composed of homogeneous individuals, concentrate on a particular area of interest are designed to extract understanding rather than information, and are more prescribed in content than a discussion. Members may agree, disagree, persuade, or argue while the facilitator probes for deeper understanding. Mini-questionnaires or handheld devices have been used to gauge individual members' ratings or agreement with subtopics of interest.

The focus group technique has been used successfully in a number of studies in diverse fields of study (Cameron, 2005; Zuckerman-Parker & Shank, 2008). Several other studies have included focus groups in research with minority students (Guiffrida, 2005; Russell, 1991; Solórzano, Ceja, & Yosso, 2000). Kao and Tienda (1998) used focus groups in conjunction with quantitative data to study the educational aspirations of minority youth. Van Aken, Watford, and Medina-Borja (1999) employed focus groups in Virginia Tech's engineering programs to assess the problems, issues, and experiences of women and minority students. They found more gender than racial problems and a need to improve the interactions and communications among minorities. Fleming (2012a) included focus groups in addition to interviews and questionnaires to assess the City College School of Engineering's CREST Center for Mesoscopic Modeling and Simulation. This Center designed an interdisciplinary research center that successfully engaged minority students in cutting edge research while also creating a friendly atmosphere in an otherwise chilly climate. Students reported being able to pursue their love of science, but worried about the lack of tenure opportunities for minorities in engineering.

Thus, this small group technique was employed to initiate student level analyses of minorities in engineering. No hypotheses were generated.

Method

Participants

A total of 176 minority engineering students participated in focus group sessions of 6–12 students in response to invitations by their program coordinators. All interviews were recorded by voice recording, video recording, or both. A population breakdown was available for the 144 students who filled out mini-questionnaires. Fifty-eight percent (84) were males, 41.7% (60) females, 51.4% (74) African American, 36.8% (53) Hispanic, and 11.8% (17) "Other" (see Table 5.1).

Focus Group Questions

The questions fell into the following nine categories—early success factors, pre-college program experience, summer bridge, minority status, program participation, faculty, job experience, career outlook, and advice to other minority students in engineering. Because focus group sessions may be dominated by one or more stronger personalities, a mini-questionnaire was administered to record the sentiments of all participants on a number of important questions. The results of the mini-questionnaire are reported in Chapter 5.

Procedure

Program coordinators in the target universities were alerted early in the spring semester that the project requested their participation in focus group sessions. The request was for two groups of 6–8 students each and one group of faculty and staff. It was left to the coordinators to pass the invitation on to students and staff. Dates for the interviews were arranged during the spring semester from February through April. All sessions were conducted by the Principal Investigator. Pizza or sandwiches were provided for students as incentives for participation. In a few cases the universities provided meals for participating students. A Focus Group Consent Form was given to all participants, and they were asked to sign their names if they consented to being recorded. The Consent Form also asked participants to indicate their classification, gender and ethnicity. Faculty/staff were asked to indicate their position in the university.

Data Analysis

Formal statistical analysis was not employed for the analysis of focus group interviews. Instead, the procedure was to describe the most common responses to the questions asked, supported by direct quotes from students.

Results

Early Exposure

How did you Become Interested in Engineering?

Students most often lead with: "I like to build things, make things, take things apart—and sometimes put them back together;" "I would try to fix everything;" "I like to build things, you name it, I like to build it." Such students began talking with their hands as if envisioning the making or breaking of things. Female students rarely made this comment, but there were two exceptions. One said she had a brother close to her age and together they would build things and make things, and sometimes tear them apart to see what made them tick. Another said she worked in her father's workshop and learned to build and make things.

Second was family influence: "My father is an engineer;" "My aunt is an engineer;" "My family is full of engineers, so I had no choice;" "I'm West Indian and expected to be a professional—I chose engineering;" "My father is in construction;" "I wanted to know why that bridge was standing up, and my father explained about expansion joints."

Third was the love of math or auspicious exposure to math: "I'm good at math;" "I love math;" "I like math and engineering lets me use it;" "I had a pretty good math teacher." In these cases, something akin to ecstasy passed over their faces at the recollection.

Fourth was program exposure. A smaller number indicated that they had participated in a program or activity that sparked the interest in engineering such as a robotics competition, or a program to interest women in science.

Fifth were the opportunities that engineering affords: "My main incentive was to come to college and get a better life;" "I had an internship when I was 18; the engineer made three times as much money as everyone else."

It seemed clear that the interests that paved the way to engineering emerged early sometimes facilitated by family or program exposure. Apart from discovering the early love of mathematics, female students were more likely to develop their interests later than males, in STEM programs, usually in high school. There is a concern in the literature that 60% of the doctorates in engineering now go to foreign nationals. It does seem from talking with these students that there is a good explanation for this fact. The largest group of students gravitate to engineering because they are doers, builders who want to get to work building their bridges and working with their hands. Most want to do this as soon as they can after earning a B.A. One student said that he didn't want a doctorate because he wanted to work outside and didn't want to get trapped in teaching. The exceptions to this rule were students in research-oriented programs where they get extensive experience in laboratories with their professors. A key question is why more foreign nationals than American students desire doctorates?

About Mathematics

It is often said, students were told, that math is the limiting factor in engineering, the skill without which engineering education cannot proceed. Do you agree with this statement, and what is your relationship to mathematics?

This question produced a flurry of strong opinions that ranged from very positive to almost negative: "Math rules;" "You have to love it;" "It's like a language that we communicate in;" "Knowing I'm good at something that everyone shies away from—is good;" "If you don't love math, you probably shouldn't be an engineer;" "It's a tool we use in all the sciences;" "Behind the wall of engineering—is math;" "You can't get away from it"; "I was originally bad at it; now I like it"; "It's a love-hate thing;" "Math is like an old girlfriend—just accept the fact that she will be in your life forever." Some students reiterated that they were very good at math and beamed with quiet pride. Others admitted to struggling with math since math in engineering is harder than high school math, and they have to take four years of it. Still others tried to explain that it isn't about the math; it's about the problem solving, and math helps you solve problems. In the final analysis, these students were clear about one thing: "No matter how we feel about math, we all do it." Is math a limiting factor in engineering? According to these students, it is, indeed.

When asked if they were computer whizzes, most denied this, but expressed comfort with them as is true of most students of their generation. "We can operate them better than your average Joe."

Parental Influence

Some students had touched on the influence of their parents in deciding on engineering, but they were asked specifically how their parents influenced their engineering education.

A good proportion had been positively encouraged toward engineering because either one of their parents, or one or more family members, were already engineers and acted as a positive force. Other families asked only that they do something professional—an attitude described most clearly by West Indian students. Students from Texas and specifically Hispanic students had fathers and other family members engaged in construction; therefore, they wanted to own construction companies or at least be the engineering supervisor who made the most money. But many other students had parents who were not in a position to encourage their engineering education; these students had single mothers or parents with very little education or parents with no idea of the possibilities in engineering. These parents, however, were all described as supportive of whatever they wanted to do. There was not one report of a non-supportive parent: "My father told me about being a Bill Gates Millennium Scholar;" "My parents pushed me toward STEM—to make more money;" "My parents were big on grades."

Program Exposure

Secondary School Exposure

Participants were asked if they had participated in STEM programs, conducted in the summer, on weekends, or as part of the school offerings, during elementary school, middle school, and high school years. Virtually none of these students had STEM program exposure during the elementary school years, and precious few in the middle school years. By high school, a significant number—perhaps close to one-third—had some form of program experience. In these cases, the programs served to develop or catalyze an interest in math, science, or engineering: "I participated in a FIRST Robotics competition and it was awesome;" "I went to a four-week summer program for women in science and discovered how good I was at math;" "I was in a program oriented to graphics design, but you could take engineering courses;" "This school offered a summer outreach program that I entered; I didn't learn that much but the people were great and encouraged me."

Summer Bridge

When asked about summer bridge programs before entering engineering school, about 40% had such experience. While most said the programs served to sharpen their skills, especially in math, almost all students were more impressed by the valuable networking they were able to do: "I advanced in math, but the networking was the best;" "I made close friends that I've stayed with;" "It taught me that networking is everything."

Educational Experiences in Engineering School

Minority Status

Students were asked, "Now that you are underrepresented minority students in engineering, is minority status an issue for you in any way?"

This was a question that many students didn't seem to know how to answer. It was also clear that the more minority-serving the institution, the less this question seemed to compute for them. Finally, many students, especially Hispanic students, shrugged and said: "Yeah, we get money. The scholarships are great;" "I find it to be a great advantage—I get two scholarships;" "We get attention from our program directors who help us a lot;" "It's hard to say since everyone here is some kind of minority—there's really no predominant group here;" "Most of our professors are not American; they are minorities, too."

African American students were the most aware of their minority status and offered the most thoughtful statements on the issue: "Yeah, it's an issue, but it just makes me work harder;" "Two professors told me that I wasn't going to make it in engineering—I had to prove them wrong;" "During my internship, I was the only Black, so I knew I had to be good;" "There's no issue here, but we know it's out there. I'll deal with it." In short, these students as a group were not bothered or burdened by their minority status, but when it became an issue, they approached it as any other problem to solve—they are problem solvers.

Being a Female in Engineering

Just as Van Aken et al. (1999) found at Virginia Tech, there appeared to be more issues surrounding female status than minority status and the women were far more vocal in describing their experiences: "I'm in computer science, which doesn't have many women, and I'm almost always the only one in class. It's weird;" "No one pays attention to what I say. A male can say the exact same thing that I did, but he is taken seriously;" "People snicker in class when I speak because I'm only a girl;" "I find myself falling into the secretarial role in our teams. It just happens, and I don't do anything about it." Even so, these women were not overly upset about their experiences. They were happy to vent, but were also happy to deal with the situation as it is. They wanted it known that there was sexism, but didn't want to belabor it. Some women said they liked being around so many men.

Distinguishing Traits

Students were asked if there were any traits or personality characteristics that distinguished them from students in other disciplines. As a probe question, students were told that the typical engineering student was traditionally described as introverted, and asked if that seemed true to them.

The question brought a smile to many of their faces as they contemplated how they might be different from non-engineering students. Clearly that contemplation also brought with it a sense of personal pride: "I think we are different; we're focused; we're in a challenging field and we just focus on that;" "Our old friends want us to go off and party, but we have studying to do, and we do it;" "Some of us are quiet, but not with each other. With each other we have a lot to talk about and we help each other;" "We work in groups, so no one has a chance to be introverted."

Thus, the general tenor of their answers suggested that they are not introverted among those with whom they share a common language (of math?) and common problems to solve. They might well appear quiet to those students who are not science oriented, or who spend time socializing rather than studying.

Program Engagement

When asked what MEP program components or activities they actively participated in, a surprising number indicated student professional organizations. They may have received tutoring or been in classes where Supplemental Instruction was offered, but it was the organizations that brought forth the most enthusiasm (note that students most likely to participate in the focus groups were the most involved): "NSBE has really helped me; the people I've met made a difference;" "I'm president of the Society of Hispanic Engineers, and this has been the best thing for me;" "The best thing about the professional organizations is the networking."

Students often sought tutoring, but felt that study groups were also invaluable. "There's less (tutoring) help for upper division courses; then you need your study group; we think things through together."

Faculty

When asked about their faculty, whether they liked them or not, whether they were good, and whether they were helpful, the reviews were mixed: "They are amazing;" "Some are OK, some aren't;" "Some of the faculty are hard to get to know and you have to be the aggressor;" "Each teacher is different;" "I get most of my help here at the Center;" "I never met one that did not try to help me during office hours." In fact, most of the engineering faculties interviewed were of foreign extraction, including Indian and Asian individuals.

Research Experience

The most positive comments about faculty came from students who had had undergraduate research experiences. In these cases, it was much more likely that student had developed a mentoring relation with them, or with a graduate student in their laboratories. However, undergraduate research experiences were not particularly common, except at specifically research-oriented universities.

Internship Experiences and Career Outlook

Internship experiences were far more common than research experience, but both provide practical experience that allows students to touch and feel the work of engineering: "In research I got to see and feel engineering, and I like to touch and feel;" "In my internship, they treat you like an engineer." Students in most schools were not guaranteed an internship. In schools neighboring industries such as Boeing or oil and gas, internships and job prospects were plentiful: "There is a company here every week." But in schools less well situated, a great deal depends on the student's GPA. This was true for both internships and job opportunities. Most students planned to go to work immediately after graduating. They wanted to begin using their skills and earning the good salaries they expected. Several students planned to return for at least a master's degree later and were actively looking for employers who would help pay for further education. A minority of students planned to continue to a doctorate, and the vast majority of these students were in research-oriented universities. There were students who planned to own their own business or become CEOs, and so planned to get an MBA: "A CEO of somebody's company is my goal." A relevant statistic is that 33% of the CEOs of Fortune 500 companies are engineers by training (Aquino, 2011). For virtually all students, the future is something they look forward to with optimism: "There's a bright future after graduation;" "I'm ready to do bigger things with my life."

Advice to Others

In one other study of advice by minority students in the health sciences, the most frequent advice for students coming after them was good time management (Fleming, 2012b). When asked this question, minority engineering students were most likely to advise "getting involved." Other advice included "develop your math skills," "manage your time," and "network." It is the emphasis on networking and interpersonal involvement that might seem surprising, in light of the stereotype of the introverted engineer. While in high school, they advise studying hard, and figuring out how to study: "High school does prepare you for college if you do high school right." Keeping the right company was another prominent theme: "Humanities students party all the time. Remember that you are in engineering and make the right friends." Persistence figured prominently in their comments: "Keep moving forward;" "Know how to cope with failure; it will happen;" "Some students had to take a course three times—now they are graduates;" "It's a matter of persevering;" "It's not easy, but you get used to it."

Conclusion

From focus group conversations, the 176 vocal and involved student participants suggested that important success factors were skill in math, which overlaps with problem-solving skill, group work, and networking. Their conversations revealed them to be the doers, problem solvers, and tinkerers who want to get out into the real world and have an impact on it. They see engineering as a pathway to a better life for themselves and those around them. While previous research has focused on the impersonal, introverted personality of the engineer, it has overlooked their pro-active approach to the world. In these qualities, minority engineering students do not seem unique or different from others—but do seem different form many minority students in treating race and ethnicity as nothing more than another problem to solve.

References

Aquino, J. (2011). 33% of CEOs majored in engineering—and the surprising facts about your boss. *Business Insider*, March 23. Retrieved from www.businessinsider.com/ceos-majored-in-engineering-2011-3.

Boddy, C. (2005). A rose by any other name may smell as sweet but 'group discussion' is not another name for a 'focus group' nor should it be. *Qualitative Market Research: An International Journal, 8*(3), 248–255.

Broberg, H., Griggs, K., & I-Hai Lin, P. (2006). Learning styles of electrical and computer engineering technology students. *Journal of Engineering Technology, 23*(1), 40–46.

Byers, P. Y., & Wilcox, J. R. (1988). *Focus groups: An alternative method of gathering qualitative data in communication research.* Paper presented at the 74th Annual Meeting of the Speech Communication Association, New Orleans, LA, November 3–6. ERIC Document No. ED297393.

Cameron, J. (2005). Focusing on the focus group. In I. Hay (Ed.), *Qualitative research methods in human geography* (2nd ed., pp. 156–174). Melbourne, Australia: Oxford University Press.

Felder, R. M., & Silverman, L. K. (1988). Learning and teaching styles in engineering education. *Engineering Education, 78*(7), 674–681.

Fleming, J. (2012a). Preparing minority students to compete at the cutting edge: The CCNY CREST Center. In J. Fleming, *Enhancing minority students retention and academic performance: What we can learn from program evaluations* (pp. 102–119). San Francisco: Jossey-Bass.

Fleming, J. (2012b). What successful students in science know about learning: Gateway to higher education. In J. Fleming, *Enhancing minority students retention and academic: What we can learn from program evaluations* (pp. 51–61). San Francisco: Jossey-Bass.

Godfrey, E., & Parker, L. (2010). Mapping the cultural landscape in engineering education. *Journal of Engineering Education, 99*(1), 5–22.

Guiffrida, D. A. (2005). Other mothering as a framework for understanding African American students' definitions of student-centered faculty. *Journal of Higher Education, 76*(6), 701–723.

Harrison, R., Tomblen, D. T., & Jackson, T. A. (2006). Profile of the mechanical engineer III. *Personnel Psychology, 8*(4), 469–490.

Heiskanen, E., Jarvela, K., Pulliainen, A., Saastamoinen, M., & Timonen, P. (2008). Qualitative research and consumer policy: Focus group discussions as a form of consumer participation. *The Qualitative Report, 13*(2), 152–172.

Kao, G., & Tienda, M. (1998). Educational aspirations of minority youth. *American Journal of Education, 106*(3), 349–384.

Krueger, R. A. (1994). *Focus groups: A practical guide for applied research.* Thousand Oaks, CA: Sage.

Morgan, D. L. (1993). *Successful focus groups: Advancing the state of the art.* Thousand Oaks, CA: Sage.

Russell, L. A. (1991). Assessing campus racism: The use of focus groups. *Journal of College Student Development, 32*(3), 271–272.

Scissons, E. H. (1979). Profiles of ability: Characteristics of Canadian engineers. *Engineering Education, 69*(8), 822–826.

Solórzano, D., Ceja, M., & Yosso, T. (2000). Critical race theory, racial micro-aggressions, and campus racial climate: The experiences of African American college students. *Journal of Negro Education, 69*(1/2), 60–73.

Stage, F. K. (Ed.). (1992). *Diverse methods for research and assessment of college students.* Alexandria, VA: American College Personnel Association.

Stewart, D. W., Shandasani, P. N., & Rook, D. W. (2006). *Focus groups: Theory and practice.* Thousand Oaks, CA: Sage.

Van Aken, E. M., Watford, B., & Medina-Borja, A. (1999). The use of focus groups for minority engineering program assessment. *Journal of Engineering Education, 88*(3), 333–343.

Zuckerman-Parker, M., & Shank, G. (2008). The town hall focus group: A new format for qualitative research methods. *The Qualitative Report, 13*(4), 630–635.

Zywno, M. S., & Waalen, J. K. (2002). The effect of individual learning styles on student outcomes in technology-enabled education. *Global Journal of Engineering Education, 6*(1), 35–44.

5 Profile of Minorities in Engineering

Analysis of Focus Group Mini-Surveys by Gender Within Ethnicity

Summary

Focus group conversations were previously reported for minority students from 11 engineering schools. This study describes the analysis of a mini-survey administered during the focus group sessions to augment group discussions which can be dominated by strong personalities, thereby skewing the results. One hundred and forty-four students, comprised of 51.4% African American, 36.8% Hispanic, and 11.8% of "Other" ethnicities, completed three open-ended questions and a fourth question on the effectiveness of seven academic support program components. Student responses were coded for thematic content, tabulated, and then entered into regression equations against four measures of achievement. Responses positively associated with achievement indices were then factor analyzed to isolate common clusters associated with success in engineering. While the most frequent student responses to the four questions were skill in math, dedication, focus, and study groups respectively, the five emerging factors associated with greater student success were: (1) participation in effective program components that provide practical engineering experience—project or problem-based courses, research experience, and industry internships; (2) a burning desire to become an engineer and reap its economic benefits; (3) taking advantage of all resources provided; (4) combining motivation and dedication with effective time management; and (5) involvement in MEP programs. The results are noteworthy in suggesting that the first and foremost success factor for minorities in engineering is exposure to engineering itself. Analyses by gender within ethnicity also suggest that success factors minority women also include inner qualities of psychological reliance and a fierce inner drive that may serve them well in the male world of engineering.

Introduction

Despite four decades of considerable effort on the part of corporations, universities, and funding agencies, minorities are still underrepresented in engineering (Chubin & Babco, 2003; Chubin, Donaldson, Olds, & Fleming, 2008). This is not to say that there has been no progress. There has. Currently, underrepresented minorities—that is, African Americans, Hispanics, and Native Americans—constitute 25% of the population but only 12% of the engineering degrees. This is up from 5% of the engineering degrees in the 1980s, according to the National Action Council for Minorities in Engineering (NACME, 2014a, 2014b). However, so modest is the progress, given the effort, that Slaton (2011) questions whether minority underrepresentation is the simple result of the historical absence of minorities in engineering, or whether it is also due to a more intractable resistance to minority inclusion in the profession, including notions that greater diversity means lower standards. Whatever the reason, she calls for studies that show which sorts of programs, program components, and features of engineering education can yield greater minority participation.

This analysis seeks to enhance a literature that barely touches on the academic and social adjustment of minorities in engineering. Whereas the chilly climate endured by women in engineering has yielded a number of studies documenting their uncomfortable position in a conservative male bastion that can serve to reduce their commitment to engineering despite their greater persistence (Borrego, Padilla, Zhang, Ohland, & Anderson, 2005; Brawner, Camacho, Lord, Long, & Ohland, 2012; Clewell, de-Cohen, Tsui, & Deterding, 2006; Espinosa, 2008; Gunter & Stambach, 2005; Lord et al., 2009; Malcolm, 2008; National Science Foundation, 2011; Rinehart & Watson, 1998), commensurate attention to the adjustment of underrepresented minorities is still to be attained. Lewis (2003) has also lamented the small number of empirical reports for minorities in science and engineering. Studies specific to minority students in engineering are limited, but do address their high dropout rates (Morning & Fleming, 1994), the negative impact of the perception of racism on retention (Brown, Morning, & Watkins, 2005), more frequent departures due to feelings of not belonging in engineering (Marra, Rodgers, & Shen, 2012), lower ratings of inclusiveness in the engineering environment (Lee, Matusovich, & Brown, 2014), reduced intellectual development (i.e., critical thinking) in White compared to Black engineering schools (Fleming, Garcia, & Morning, 1996), lack of support from instructors beginning in high school (Fleming & Morning, 1998), and lower ratings of abilities compared to non-minorities in engineering (Ro & Loya, 2015). There are also studies describing programs and program features that work (Lam, Srivatsan, Doverspike, Vesalo, & Mawasha, 2005; Landis, 2005; Maton, Watkins-Lewis, Beason, & Hrabowski, 2015; Morning & Fleming, 1994; Fleming, 2012a, 2012b; Mitchell, 2007; Mitchell & Daniel, 2004; Reichert, 1997; Slaton, 2011; Swail & Chubin, 2007; Yelamarthi & Mawasha, 2008).

The broad study employs multiple methodologies, ranging from institution-level statistical analyses and the impressions gathered during focus group discussions to quantitative analyses of academic performance and adjustment. Chapter 4 described the impressions obtained from multiple focus group conversations at eleven selected universities. They described a group of students highly engaged in their program activities and students who gravitated to engineering largely by inclination or family influence, were groomed by exposure to STEM programs in secondary school and summer bridge, who thrive in group work and student organizations, who are inclined to solve problems rather than to dwell on setbacks of any kind and therefore are in position to have an impact on the world. To augment these impressions, the present study examined the mini-questionnaire administered during the focus group sessions.

The focus group method often employs multiple methodologies or is itself one of a number of data-gathering strategies that have been employed to study minority students (Guiffrida, 2005; Kao & Tienda, 1998). Focus groups have traditionally been used in marketing research (Eysenbach & Köhler, 2002; Griffin & Hauser, 1993), but have found a niche in engineering education research and evaluation (Litchfield & Javernick-Will, 2015; Kontio, Lehtola, & Bragge, 2004; Martınez, Dimitriadis, Rubia, Gómez, & De La Fuente, 2003; Mawdesley, Long, Al-Jibouri, & Scott, 2011; Natishan, Schmidt, & Mead, 2000; Seymour & Hewitt, 1997). Fleming (2012b) conducted focus groups with minority engineering students engaged with faculty in research enterprises, but also administered a mini-questionnaire to ensure input from each student. Focus groups, however, have their limitations, as Griffin and Hauser (1993) and Kontio et al. (2004) point out. The group mind may obscure important individual differences as dominant personalities control the discussion. In the present study, focus group discussions were supplemented by a four-question mini-survey to ensure sure that each respondent could weigh in on some of the critical issues related to success factors for minorities in engineering. The analysis of the mini-survey is the subject of this chapter. Further, while the focus group interviews permitted little opportunity to explore gender or ethnic differences, the mini-survey does allow a consideration of overall success factors but also those that may also be unique to Black and Hispanic males and females.

A mini-questionnaire was administered to record the sentiments of 144 participants on several important questions. The first three were open-ended; the last required students to rate their level of agreement with the effectiveness of seven program components. The responses obtained were submitted to a series of statistical procedures to reach conclusions as to the pathways of successful minorities in engineering. Table 5.1 presents a breakdown of the subject population, and Appendix 5.1 contains the details of method.

Table 5.1 Focus Group Population

Group	Students
All	n = 144
Males	58.3%
	n = 84
Females	41.7%
	n = 60
African American	51.4%
	n = 74
African American Males	51.4%
	n = 38
African American Females	48.6%
	n = 36
Hispanic	36.8%
	n = 53
Hispanic Males	66.0%
	n = 35
Hispanic Females	34.0%
	n = 18
Other: Native American, International	11.8%
	n = 17
Other Males	64.7%
	n = 11
Other Females	35.3%
	n = 6

Note. Ethnic breakdowns within gender use the gender group as the percentage referent.

Results

What Motivated Students to Become Engineers?

As shown in Figure 5.1, the top three reasons why minority students say they became interested in engineering were:

1. Math skill, endorsed by 38.2%: "I disliked all subjects other than math;" "I have always loved math;" "Teachers said I should because I liked math."
2. Love of science, according to 34.0%: "I was very much interested in science;" "I loved science from an early age;" "In high school, I was very interested in science which was influenced by some amazing science teachers."
3. Family influence and support, by 20.8%: "My dad is an engineer;" "My family is predominantly engineers;" "My parents are engineers."

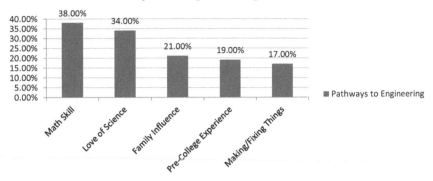

Figure 5.1 How Did You Become Interested in Engineering? Top Responses of Minority Engineering Students

This pattern of motivations was largely true of all students, regardless of gender or ethnicity, with a few exceptions. Hispanic females were the most likely to cite math skill as their primary motivation (61.1%), and they were significantly more likely than Hispanic males to so indicate (28.6%). Also, female students, especially Hispanic females, were more likely to cite love of science as their prime motivation. Female students were more likely than males to cite family influence and support (30.0% vs. 14.3%).

Other frequent sources of interest in engineering were pre-college experience in math and science programs (18.8%), inclination toward making and fixing things (17.4%), for the opportunities engineering affords for career and for society (14.6%), desires to improve the world and help people (10.4%), and love of computers and electronics (10.4%). Male students were more likely than females to cite love of computers and electronics as their entryway into engineering (15.5% vs. 3.3%), and the gender disparity was significant among African American students (21.1% vs. 2.78%). Other sources of interest fell below 10% in frequency. Interestingly, the inclination toward making and fixing things, which was prominent in focus group sessions, was only committed to writing in 17% of the cases.

So, what do students with the highest GPAs say motivated them to become engineers, and are their motivations different from the group responses? Appendix 5.2 displays the regression analysis and shows that higher GPAs were most associated with choosing engineering for the opportunities it affords: more money; the options it provides; and the ability to make a better life for family. It is worth noting that when the group is disaggregated, high-performing males, especially Hispanic males, most often chose engineering for its career opportunities. Hispanic students, especially high-performing Hispanic females, choose engineering because of the inclination to make and fix things, and because of their math skills. For the small number of students of "Other" ethnicities, those with the best academic performance most often chose engineering because of their desire to improve the world.

In short, the predominate interests behind the choice of minority engineering students were, predictably, math skill, love of science and family influence. However, the highest performing students were also motivated by good career opportunities (the hunger factor), the bent toward making and fixing things, and the desire to improve the world.

Success in Engineering

The second question put to them was "What does it take to be successful in your engineering program?" The top three responses were (see Figure 5.2):

1. Dedication and motivation, from 49.3% of students: "Dedication and motivation because there is a lot to learn and a lot to do;" "Commitment;" "Determination."
2. Effort and hard work, from 31.3%: "It takes a very strong work ethic;" "Work on homework and study almost every day;" "Work hard."
3. Networking and working well with others, from 29.9%: "You have to know how to network;" "Being able to work in groups and hear other people's ideas;" "Network with other students of color and other students in general."

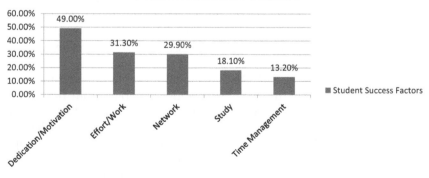

Figure 5.2 What Does It Take to Succeed in Your Engineering Program?

Again, students were in general agreement on the most important ingredients of success in engineering school. African Americans, however, more often cited effort and hard work than did Hispanic students (41.9% vs. 18.9%). African American males, more than their female counterparts, cited effort and hard work (55.3% vs. 27.8%).

When the various ingredients for success offered by respondents were entered into a regression equation to determine those ingredients associated with higher student GPA, surprisingly few were positively related to academic success (Appendix 5.3). More were negatively related, as were money (scholarships) and discipline. Among female students, networking and program involvement (as in MEP programs) were more often mentioned by high-performing women. Among high-performing Hispanic females, the desire to be an engineering was mentioned most. Finally, among students of "Other" ethnicities, study, dedication and motivation, and use of the 4.0 Learning system were more often cited by high performers.

In short, there is general consensus on the ingredients for success in engineering school: dedication/motivation; effort/hard work; and networking. When high-performing students differed from other students, they cited MEP program involvement, the strong desire to be an engineer, the 4.0 Learning system, and study, study, study as necessary success ingredients.

Advice to New Students

Figure 5.3 displays the advice that respondents would give to other minority students entering engineering:

1. To prioritize and focus, 31.0%: "Focus on end goal after 4 years;" "They must keep their eye on the goal they want to accomplish;" "Stay focused on your goal."
2. To network, 30.0%: "Network, network, network;" "Make sure you build great relationships;" "Find study buddies."
3. Dedication/motivation/perseverance, 18.0%: "If your motivation is lost, find it again—quickly;" "Be prepared for successes and failure, but never give up;" "Persevere." There was general agreement among student groups, and no significant differences between them.

As shown in Appendix 5.4, the advice significantly related to higher GPA in a regression equation was dedication and motivation. This was especially true among male students, and Hispanic students. High-performing students of "Other" ethnicities advised campus involvement. High-performing African American females advised getting involved, and networking.

In short, while minority engineering students in general advise incoming minorities to focus, network, and be dedicated, high-performing

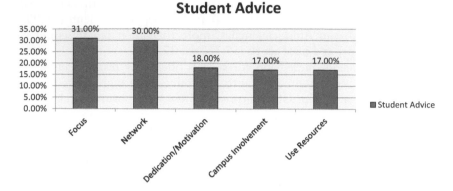

Figure 5.3 Advice to Incoming Minority Engineering Students

students also advised campus involvement in addition to dedication and networking. Thus, there seems to be little real difference in the group advice and advice from better performing students.

Program Participation

Students were asked to indicate their assessment of the degree of effectiveness of seven generic program components. Figure 5.4 shows that the highest effectiveness ratings were given to study groups (4.65 of 6.00), followed by project or problem-based courses (4.25) and tutoring (4.19). There were no significant differences among student groups in these assessments.

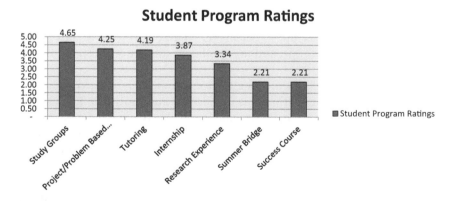

Figure 5.4 Program Component Ratings

Again, a number of the program component assessments were negatively correlated with academic achievement, as shown in Appendix 5.5. However, high-performing students gave the highest ratings to project or problem-based courses. High-performing female students gave the highest effectiveness ratings to industry internships, and this was especially true of Hispanic females.

Thus, academic support programs received relatively high marks, yet the highest-performing students reserve higher ratings for program components that provide more hands-on, practical engineering experiences, as in problem or project-oriented coursework or real industry experience gained through internships.

Success Factors

Thus far, we have correlated student responses only with GPA to determine what sentiments might be associated with academic success. The drawbacks to this method are twofold. The measure of GPA relies on reported scores, which may be less than wholly accurate, despite their widespread use in education research (Kuncel, Crede, & Thomas, 2005). Second, these scores were obtained across a number of institutions, which may blur their significance. Thus, we considered three additional measures of success: longevity or retention in the program (i.e., the student's classification); the average SAT/ACT scores of the minority students in the school of matriculation, a measure of institutional success or prestige; and the success of the institution in graduating minorities over a six-year period, another measure of institutional success.

Table 5.2 presents the regressed correlates with all four measures of success resulting from the interest in engineering, success ingredients, advice, and program ratings for the whole sample of students. It shows that there were 12 variables positively associated with measures of success. These 12 variables were then submitted to the factor analyses presented in Table 5.3, which resulted in the isolation of five success factors. These were:

1. Factor 1: *Hands-On/Experience-Based Program Components.* These program components, judged more effective by students high-performing by multiple definitions, describe program components that provide practical engineering experience: project or problem-based courses; research experience; and industry internships.
2. Factor 2: *Desire for the Opportunities in Engineering.* This might be called the hunger factor. The burning desire to be an engineer meets the quest to seize the better earnings and promise of a better life.
3. Factor 3: *Using All Available Resources.* Both as an assessment of what it takes to succeed and as their best advice, high performers advocate taking advantage of the academic support provided for students.

Table 5.2 Student Responses Correlated With Four Measures of Achievement

Group	Variable	t-Value	Multiple R/ Variance Accounted	F-Value
With GPA				
All (n = 130) Interest in Engineering	**Opportunities**	2.07[*]	R = .181 R² = .033	F = 4.29[*]
All (n = 129) What It Takes	Scholarship	−2.54[*]		
	Discipline	−1.99[*]	R = .266 R² = .071	F = 4.83[**]
Advice: All (n = 98)	**Dedication/Motivation** Time/Management CW	2.24[*] −2.34[*]	R = .330 R² = .109	F = 5.86[**]
All (n = 124) Program Component Evaluations	Summer Bridge	−2.52[*]		
	Project/Problem-Based Courses	2.35[*]	R = .271 R² = .074	F = 4.84[**]
With Longevity:				
Interest	Robotics	−2.06[*]	R = .172 R² = .030	F = 4.25[*]
All (n = 142) What It Takes	Organization	−3.33[***]		

(Continued)

Table 5.2 (Continued)

Group	Variable	t-Value	Multiple R/ Variance Accounted	F-Value
	Use Resources	2.02^*	$R = .322$ $R^2 = .102$	8.09^{***}
All (n = 98) Advice	Plan Ahead CW	-2.39^*	$R = .235$ $R^2 = .055$	5.69^*
All (n = 124) Program Component Evaluations	Research Experience	2.93^{**}		
	Internship Success Course	3.29^{***} -2.43^*	$R = .421$ $R^2 = .177$	9.54^{***}
With Average SAT of Minorities for the School:				
All (n = 142) Interest in Engineering	Pre-College Experience Organizations Study Habits	-2.48^* -2.30^* -2.30	$R = .319$ $R^2 = .102$	5.26^{**}
All (n = 143) What It Takes	Programs (like PROMES) (Use) Resources	2.42^* 2.42^*	$R = .272$ $R^2 = .074$	5.61^{**}
All (n = 99) Advice	(Use) Resources	2.17^*	$R = .214$ $R^2 = .046$	4.71^*
All (n = 108) Program Component Evaluations	NA			

(Continued)

Table 5.2 (Continued)

Group	Variable	t-Value	Multiple R/ Variance Accounted	F-Value
With Adjusted School Rank for 6-Year Graduation Rates:				
All (n = 108) Interest in Engineering (Note: Higher ranks = schools where it is difficult for minorities to graduate)	Improve World	2.50*		
	Making, Fixing Things	2.18*	R = .288 R^2 = .083	4.79**
All (n = 108) What It Takes	**Desire**	-2.38^*		
	Time	2.29*	R = .304 R^2 = .093	5.41**
All (n = 99) Advice	**Time CW**	-2.39^*		
	Friends CW	2.03*	R = .314 R^2 = .099	5.31*
All (n = 108) Program Component Evaluations	**Internship**	-3.62^{***}	R = .336 R^2 = .113	13.09***

Note. Variables positively correlated with achievement indices in italics. "CW" denotes critical word coding, as opposed to thematic coding.

* p < .05
** p < .01
*** p < .001

Table 5.3 Factor Analysis of Student Mini-Questionnaire Reponses Positively Correlated With Four Measures of Achievement

Extracted Factors	Factor Loadings
Factor 1:	**Hands-On/Experience-Based Program Components**
Internship Evaluation	.749
Project/Problem-Based Courses Evaluation	.700
Research Experience Evaluation	.699
Factor 2:	**Desire for the Opportunities in Engineering**
Q2 Desire	.781
Q1 Opportunities	.697
Factor 3:	**Use Available Resources**
Q3 (Use) Resources	.816
Q2 (Use) Resources	.568
Factor 4:	**Dedication vs. Time Management**
Q3 Dedication/Motivation	−.768
Q3 Time CW	.665
Factor 5:	**MEP Programs**
Q2 Programs	.928

Note. "CW" denotes critical word coding, as opposed to thematic coding.

4. Factor 4: *Dedication vs. Time Management.* This bi-polar factor may describe two attitudinal approaches to success that may be complimentary rather than in opposition, and represent a semantic difference rather than a true dichotomy. One set of students thinks in terms of dedication and the motivation engineering requires, while the other thinks in terms of the time and time management skills required to reach the goal.
5. Factor 5: *MEP Programs.* This factor describes students who realize the benefit of academic support programs in engineering, which encompass much of what it takes to succeed.

If we were to summarize these factors for student consumption, we might suggest that to succeed in engineering, they would be well advised to: (1) get practical experience in what engineers actually do; (2) go with gusto after the better life that engineering provides; (3) be resourceful and take advantage of all that is provided for you; (4) find enough motivation and dedication to put in and manage the time necessary; and (5) give thanks for the MEP programs available, and get involved in them.

Success Factors by Gender Within Ethnicity

Do the success factors observed so far vary significantly by gender within ethnicity? To answer this question, Table 5.4 presents the descriptive factor-analyzed variables associated with success measures separately for the four major gender-ethnic groups: African American males and females; Hispanic males and females. It does seem that each of the groups has much in common in that positive involvements of various kinds in undergraduate engineering education seem prominent. Successful African American women, in our terms, use the word "involvement" in their responses and specifically advise networking. African American males also advise networking, and were characterized by positive involvement in three program components (internships, success courses, and problem/project-based courses). For Hispanic females, positive involvement in two program components were associated with success (research and internships), while Hispanic male success was associated with general campus-wide involvement, as well as involvement in three program components (summer bridge, success course, and project/problem courses). In short, involvement is key, whether it be networking with others, campus activities, or academic programs offered to further student development.

Yet there were discernible differences among the groups. For example, both groups of women were most characterized by the kinds of advice they gave; that is, 75% of the Black female success factors and 60% of Hispanic female success factors were advice variables, as opposed to 16.7% of Black males and 0% of Hispanic males. So, what is the advice that these two groups of minority women in engineering give? Black females advise involvement in general and networking in particular, as already discussed. However, they also advise having or cultivating the right mindset; i.e., the right kind of lifestyle, personality or "head" for engineering. They are advising a kind of psychological or internal resilience that is unique to them. Hispanic women, on the other hand, advise desire, focus, dedication, and motivation that bespeaks a fierce inner drive. Both groups of successful women, then, offer inner—or perhaps introspective—advice that seems uncharacteristic of the typical engineer. Indeed, the males of these groups avoid introspective comments. Black males were distinguished by the interests that brought them to engineering—their interest in computers and their interest in the career opportunities in engineering. Successful Hispanic males were distinguished by their success strategies—their idea of what it takes to succeed in engineering: campus involvement; discipline; and problem solving. In sum, while the gender by ethnic groups have much in common, subtle differences indicate that successful women in this group more often voice inner qualities that guide them in this man's world.

Table 5.4 Factor Analysis of Mini-Questionnaire Reponses Positively Correlated With Four Measures of Achievement by Gender Within Ethnicity

African American Female Mini-Questionnaire Reponses

	Involvements of Various Kinds
Factor 1:	
Advice: Involved CW	-.735
Program Component: Internship Evaluation	.726
Advice: Network	.592

	Personality/Mindset
Factor 2:	
Advice: Personality/Lifestyle	.851

Hispanic Female Mini-Questionnaire Reponses

	Program Involvement
Factor 1:	
Program Component: Research Experience Evaluation	.914
Program Component: Internship Evaluation	.904

	Desire/Focus or Dedication/Motivation
Factor 2:	
Advice: Desire	.757
Advice: Focus CW	.698
Advice: Dedication/Motivation	-.621

(*Continued*)

Table 5.4 (Continued)

African American Male Mini-Questionnaire Reponses

Factor 1:   *Network*
Advice: Network   .960

Factor 2:   *Program Involvement*
Program Component: Internship Evaluation   .782
Program Component: Success Course Evaluation   .704
Program Component: Project/Problem-Based   .681
     Course Evaluation

Factor 3:   *Interest via Computers or Opportunities*
Interest: Computers   .893
Interest: Opportunities   -.560

Hispanic Male Mini-Questionnaire Reponses

Factor 1:   *Campus Involvement*
Program Component: Summer Bridge Evaluation   .770
To Succeed: Campus Involvement   .758

Factor 2:   *Success Course*
Program Component: Success Course Evaluation   .921

Factor 3:   *Project Course or Discipline*
Program Component: Project/Problem-Based   .816
     Course Evaluation
To Succeed: Discipline   -.754

Factor 4:   *Problem Solving vs. Opportunities*
To Succeed: Problem Solve   .780
Interest: Opportunities   -.678

Note. "CW" denotes critical word coding, as opposed to thematic coding.

Conclusion

In contrast to focus groups discussions which are vulnerable to being dominated by strong or loquacious personalities, the addition of a mini-questionnaire allows each individual to weigh in on important aspects of the subject at hand. Indeed, in this case a different landscape emerges, compared to the focus group discussion, in response to this series of four questions, including three open-ended questions and one evaluation series.

First, how did these students come to be interested in engineering? The most usual routes were: math skill; love of science; and the influence of engineers in the family. Second, what does it take to be successful in their engineering programs? In their view: dedication and motivation; effort and hard work; networking; and working well with others. Third, their advice to other minority students entering engineering was: focus; network; and find receptive instructors. Finally, after being asked to rate the effectiveness of seven program components, students gave the highest marks to: study groups; project- or problem-based courses; and tutoring.

This group of students is, in all likelihood, a select group who are actively engaged in department activities and well known to their staff. Thus, they may not be representative of the average minority student in engineering. But of the more successful of this engaged group of students, do their opinions on our critical questions differ from those of the rest? There was far less difference of opinion between students with higher vs. lower reported GPAs than might have been expected. In some cases, the differences were more a matter of semantics rather than underlying themes. More differences emerged when student responses were correlated with four measures of achievement: GPA' longevity (i.e., classification); average test scores of fellow minority students; and the engineering school's success in graduating minorities. After being subjected to analytic procedures to isolate common factors, these suggested that success for minorities in engineering owes to: a burning desire to be an engineer and reap its rewards; dedication and time management; hands-on experience with the practical work that engineers do in solving problems, in research, and in industry; and the wisdom to use all available resources, including the embracing of MEP programs.

The average responses of involved minority engineering students to the four questions posed described individuals with math and science leanings, who know the importance of networking and studying together. However, this investigation is geared toward defining success factors rather than average factors. As such, the cluster analysis of success factors takes on primary importance. Perhaps the most revealing aspect of this analysis is the prominence of exposure to the work of engineering described in Factor 1: problem-based courses, research experience, and internships. The prominence of courses built around engineering problems or projects is consistent with recent research describing their importance in teaching engineering concepts as well as promoting retention (Lam et al., 2005; Mills & Treagust, 2003; Yadav, Subedi, Lundeberg, & Bunting, 2011). Research experiences have come to our attention as high impact retention strategies in higher education (Kuh, 2008), as well as a critical method for catalyzing interest in the sciences (Maton & Hrabowski, 2004; Maton, Domingo, Stolle-McAllister, Zimmerman, & Hrabowski, 2009), and promoting entry into engineering (Lain & Frehill, 2012). The importance of internships as a success factor could be presaged by the success of engineering co-op programs (Raelin et al., 2014), even though they are not available to all students. It is not clear from this study which came first: success in engineering that opened pathways to engineering exposures; or engineering exposures that catalyzed student success. Nonetheless, it does make sense that students with math and science leanings who have the benefit of exposure to the dazzling frontiers in the real world of engineering could easily have their career inclinations set on fire. It would seem, then that the greatest success factor for minorities in engineering is, in fact, exposure to engineering itself.

While this is the overall picture, subtle differences emerged when examining the success factors separately by gender within ethnicity. For all of the groups, involvement of various kinds in the campus environment, with others, and in critical program components defined the common experience of success. Successful Black males were most defined by their interests in computers and making the most of their career opportunities; Hispanic males by their success strategies of campus involvement, discipline, and problem solving. In subtle contrast, the success factors of minority women described inner qualities that might help explain their survival and success in the male world of engineering. For Black women, it is the "right" mindset they describe and the psychological resilience it suggests. For Hispanic women, it is their inner drive and motivation to succeed.

References

Borrego, M. J., Padilla, M. A., Zhang, G., Ohland, M. W., & Anderson, T. J. (2005). *Graduation rates, grade-point average, and changes of major of female and minority students entering engineering.* Proceedings, ASEE/IEEE Frontiers in Education Conference, October 19–22, Indianapolis, IN.

Brawner, C. E., Camacho, M. M., Lord, S. M., Long, R. A., & Ohland, M. W. (2012). Women in industrial engineering: Stereotypes, persistence, and perspectives. *Journal of Engineering Education, 101*(2), 288–318.

Brown, A. R., Morning, C., & Watkins, C. (2005). Influence of African American engineering student perceptions of campus climate on graduation rates. *Journal of Engineering Education, 94*(4), 263–271.

Chubin, D. E., & Babco, E. (2003). *"Walking the talk" in retention-to-graduation: Institutional production of minority engineers—A NACME analysis.* Retrieved from www. nacme.org/news/researchletter. html.

Chubin, D. E., Donaldson, K., Olds, B., & Fleming, L. (2008). Educating generation net: Can U.S. engineering woo and win the competition for talent? *Journal of Engineering Education, 97*(3), 245–258.

Clewell, B. C., deCohen, C. C., Tsui, L., & Deterding, N. (2006). *Revitalizing the nation's talent pool in STEM: Science, technology, engineering and math.* Washington, DC: Urban Institute.

Espinosa, L. L. (2008). The academic self-concept of African American and Latina(o) men and women in STEM majors. *Journal of Women and Minorities in Science and Engineering, 14*(2), 177–203.

Eysenbach, G., & Köhler, C. (2002). How do consumers search for and appraise health information on the world-wide web? Qualitative study using focus groups, usability tests, and in-depth interviews. *BMJ, 324*(7337), 573–577.

Fleming, J. (2012a). Retaining students in engineering at the City College of New York: How a successful program works. In J. Fleming, *Enhancing minority students retention and academic performance: What we can learn from program evaluations* (pp. 74–85). San Francisco: Jossey-Bass.

Fleming, J. (2012b). Preparing minority students to compete at the cutting edge: The CCNY CREST Center. In J. Fleming, *Enhancing minority students retention and academic performance: What we can learn from program evaluations* (pp. 102–119). San Francisco: Jossey-Bass.

Fleming, J., Garcia, N., & Morning, C. (1996). The critical thinking skills of minority engineering students: An exploratory study. *Journal of Negro Education, 64*(4), 437–453.

Fleming, J., & Morning, C. (1998). Correlates of the SAT in minority engineering students: An exploratory study. *Journal of Higher Education, 69*(1), 89–108.

Griffin, A., & Hauser, J. R. (1993). The voice of the customer. *Marketing Science, 12*(1), 1–27.

Guiffrida, D. A. (2005). Other mothering as a framework for understanding African American students' definitions of student-centered faculty. *Journal of Higher Education, 76*(6), 701–723.

Gunter, R., & Stambach, A. (2005). Differences in men and women scientists' perceptions of workplace climate. *Journal of Women and Minorities in Science and Engineering, 11*(1), 97–116.

Kao, G., & Tienda, M. (1998). Educational aspirations of minority youth. *American Journal of Education, 106*(3), 349–384.

Kontio, J., Lehtola, L., & Bragge, J. (2004, August). Using the focus group method in software engineering: Obtaining practitioner and user experiences. In *Empirical Software Engineering, 2004. ISESE'04* (pp. 271–280). New York: IEEE.

Kuh, G. (2008). *High-impact educational practices: What they are, who has access to them, and why they matter.* Washington, DC: Association of American Colleges and Universities.

Kuncel, N. R., Crede, M., & Thomas, L. L. (2005). The validity of self-reported grade point averages, class ranks, and test scores: A meta-analysis and review of the literature. *Review of Educational Research, 75*(1), 63–82.

Lain, M. A., & Frehill, L. M. (2012). *2010–2011 NACME graduating scholars survey results.* White Plains, NY: National Action Council for Minorities in Engineering.

Lam, P. C., Srivatsan, T., Doverspike, D., Vesalo, J., & Mawasha, P. R. (2005). A ten year assessment of the pre-engineering program for under-represented, low income and/or first generation college students at the University of Akron. *Journal of STEM Education, 6*(3/4), 14–19.

Landis, R. (2005). *Retention by design.* New York: National Action Council for Minorities in Engineering.

Lee, W. C., Matusovich, H. M., & Brown, P. R. (2014). Measuring underrepresented student perceptions of inclusion within engineering departments and universities. *International Journal of Engineering Education, 30*(1), 150–165.

Lewis, B. F. (2003). A critique of literature on the underrepresentation of African Americans in science: Directions for future research. *Journal of Women and Minorities in Science and Engineering, 9*(3/4), 361–373.

Litchfield, K., & Javernick-Will, A. (2015). I am an engineer AND: A mixed methods study of socially engaged engineers. *Journal of Engineering Education, 104*(4), 393–416.

Lord, L. M., Camacho, M. M., Layton, R. A., Long, R. A., Ohland, M. W., & Wasburn, M. H. (2009). Who's persisting in engineering? A comparative analysis of female and male Asian, Black, Hispanic, Native American, and White students. *Journal of Women and Minorities in Science and Engineering, 15*(2), 167–190.

Malcolm, S. (2008). The human face of engineering. *Journal of Engineering Education, 97*(3), 237–238.

Marra, R. M., Rodgers, K. A., & Shen, D. (2012). Leaving engineering: A multi-year single institution study. *Journal of Engineering Education, 101*(1), 6–27.

Martınez, A., Dimitriadis, Y., Rubia, B., Gómez, E., & De La Fuente, P. (2003). Combining qualitative evaluation and social network analysis for the study of classroom social interactions. *Computers & Education, 41*(4), 353–368.

Maton, K. I., Domingo, M. R. S., Stolle-McAllister, K. E., Zimmerman, J. L., & Hrabowski, F. A. III (2009). Enhancing the number of African Americans who pursue STEM Ph.Ds: Meyerhoff scholarship program outcomes, processes, and individual predictors. *Journal of Women and Minorities in Science and Engineering, 15*(1), 15–37.

Maton, K. I., & Hrabowski, F. A. III. (2004). Increasing the number of African American Ph.D.s in the sciences and engineering: A strengths-based approach. *American Psychologist, 59*(6), 629–654.

Maton, K. I., Watkins-Lewis, K. M., Beason, T., & Hrabowski, F. A. III. (2015). Enhancing the number of African Americans pursuing the PhD in engineering: Outcomes and processes in the Meyeroff Scholarship program. In J. B. Slaughter, Y. Tao, & W. Pearson, Jr. (Eds.), *Changing the face of engineering: The African American experience* (pp. 354–386). Baltimore, MD: Johns Hopkins University Press.

Mawdesley, M., Long, G., Al-Jibouri, S., & Scott, D. (2011). The enhancement of simulation based learning exercises through formalised reflection, focus groups and group presentation. *Computers & Education, 56*(1), 44–52.

Mills, J. E., & Treagust, D. F. (2003). Engineering education—Is problem-based or project-based learning the answer? *Australasian Journal of Engineering Education, 3*(2), 2–16.

Mitchell, T. L. (2007). START: A formal mentoring program for minority engineering freshmen. *Proceedings, American Association for Engineering Education, 2785.*

Mitchell, T. L., & Daniel, A. (2004). A model for ensuring diversity in engineering recruiting and scholarship administration. *Proceedings, International Conference on Engineering Education, Gainsville, Florida, October.*

Morning, C., & Fleming, J. (1994). Project preserve: A program to retain minorities in engineering. *Journal of Engineering Education, 83*(3), 237–242.

NACME (National Action Council for Minorities in Engineering). (2014a). African Americans in engineering. *Research & Policy Brief, 4*(1, April), 1–2.

NACME (National Action Council for Minorities in Engineering). (2014b). Latinos in engineering. *Research & Policy Brief, 4*(3, October), 1–2. August), 1–2.

National Science Foundation. (2011). *Women, minorities and persons with disabilities in science and engineering.* Arlington, VA: Author. Retrieved from www.nsf.gov/statistic.

Natishan, M. E., Schmidt, L. C., & Mead, P. (2000). Student focus group results on student team performance issues. *Journal of Engineering Education, 89*(3), 269–272.

Raelin, J. A., Bailey, M. B., Hamann, J., Pendleton, L. K., Reisberg, R., & Whitman, D. L. (2014). The gendered effect of cooperative education, contextual support, and self-efficacy on undergraduate retention. *Journal of Engineering Education, 103*(4), 599–624.

Reichert, M., & Absher, M. (1997). Taking another look at educating African American engineers: The importance of undergraduate retention. *Journal of Engineering Education, 86*(3), 241–253.

Rinehart, J., & Watson, K. (1998). *A campus climate survey at Texas A&M University.* Proceedings of the 1998 women in engineering conference: Creating a global engineering community through partnerships, 93–100, West Lafayette, IN.

Ro, H. K., & Loya, K. I. (2015). The effect of gender and race intersectionality on student learning outcomes in engineering. *The Review of Higher Education, 38*(3), 359–396. doi: 10.1353/rhe.2015.0014.

Seymour, E. & Hewitt, N. (1997). *Talking about leaving: Why undergraduates leave the sciences.* Boulder, CO: Westview Press.

Slaton, A. (2011). *Race, rigor, and selectivity in U.S. engineering: The history of an occupational color line.* Cambridge, MA: Harvard University Press.

Swail, W. S., & Chubin, D. E. (2007). *An evaluation of the NACME block grant program.* Virginia Beach, VA: Educational Policy Institute.

Yadav, A., Subedi, D., Lundeberg, M. A., & Bunting, C. F. (2011). Problem-based learning: Influence on students' learning in an electrical engineering course. *Journal of Engineering Education, 100*(2), 253–280. Baltimore, MD: Johns Hopkins University Press.

Yelamarthi, K., & Mawasha, P. R. (2008). A pre-engineering program for the under-represented, low-income and/or first-generation college students to pursue higher education. *Journal of STEM Education: Innovations and Research*, *9*(3/4), 5–15.

Appendix 5.1
Methodological Details

Participants

The participants were 144 students who filled out the mini-questionnaire. Table 5.1 presents the gender and ethnic breakdown of this group. It shows that students were composed of 58.3% males and 41.7% females. Their ethnic composition was 51.4% African American, 36.8% Hispanic, and 11.8% of "Other" groups including Native American and International students.

The focus group constituents were recruited by university program liaisons. They were asked to assemble 6–8 students in each of two student sessions and one faculty/staff session. Thus, involved students known to their staff were the most likely to participate. The Principal Investigator conducted all of the focus group sessions. Participants came from the following universities, chosen because of their cooperation in the prior statistical study of institutions with MEPs (Minority Engineering Programs). These institutions, listed in order of visitation, were: Prairie View A&M; University of Texas, San Antonio; University of Houston; Kettering Institute; Georgia Tech; University of Washington; Virginia Tech; University of Central Florida; North Carolina A&T; City College of New York; and University of California, San Diego. The first three universities represented trial runs, whereupon adjustments were made in the procedure. An additional question was added to the mini-questionnaire, and students were asked to report their GPA. Thus, the numbers of students in some analyses vary.

Measures

The mini-questionnaire asked the following open-ended questions:

1. How did you become interested in engineering? What motivated you to become an engineer?
2. What does it take to be successful in your engineering program?
3. What advice would you give to students coming after you?

The final question asked for students' ratings of program components on a Likert-type scale ranging from 0 (no experience or familiarity), and 1 (very ineffective) to 6 (very effective) for the following programs: Summer Bridge; Freshman Orientation "Success Course" for Engineering Students; Tutoring; Project-Based or Problem-Based Course(s); Study Groups; Faculty-Led Undergraduate Research Experience; Industry Internship Program.

Note that the third open-end question on "advice" was not added until the third institution, as was the measure of reported GPA.

In order to assess the "efficacy" of the students responses given, statistical procedures were used to determine their degree of association with four measures of "success," two individual measures, and two institutional measures: (1) From a 12-category checklist, students indicated their approximate GPA; (2) student classification was used to assess longevity in the program or as an informal measure of retention; (3) the average minority student SAT/ACT score, determined from the previous statistical study; and (4) the institution's rank-ordered success in graduating minority students, also determined from the previous statistical study.

The analysis describes the most frequent student responses, as well as responses as a function of GPA. For the remaining three measures of success, only the overall summative analysis was presented here and used in the effort to isolate "success factors."

Analysis

A content analysis was conducted from the written answers to the three open-ended questions. The responses were read by two individuals, who then devised categories of responses and reexamined the responses for coding purposes. Three methods of extracting themes were used: (1) coding based on thematic similarity, which may require an inference by the coder; e.g., the category "passion" defined by "this must be your passion;" "love what you do;" (2) coding based on the use of a critical word, such as "time," "schedule," or "faculty/instructor/teacher;" (3) identification of themes differentiated by high or low academic performance; in this case, the differences in responses from two or more split groups were used as the basis for extracting a theme.

Descriptive statistics (percentages) were used to report the prominent themes. Chi Square analysis was used to determine significant differences in the frequency of responses between gender and ethnic groups; in all cases, the Fisher Exact Test results were selected. For correlation-based analyses (i.e., regression and factor analyses), the categorical responses (present, absent) extracted were converted to orthogonal coding (+1, −1) to facilitate the base correlation matrix.

Regression analyses were then used to determine responses associated with measures of individual and institutional success. Finally, factor analyses were used to isolate common factors across the four measures of success. The resulting factors were used solely for descriptive purposes.

Appendix 5.2

Student Responses Associated With Academic Performance (GPA): How Did You Become Interested in Engineering?

Group	Variable	t-Value	Multiple R/ Variance Accounted For	F-Value
All (n = 130)	*Opportunities*	2.07*	R = .181 R² = .033	4.29*
Male (n = 74)	Math Skill	−3.35***		
	Opportunities	2.50*	R = .437 R² = .191	8.51***
Females	NA	NA	NA	NA
African American (n = 64)	Exposure/Person	NA	NA	NA
Hispanic (n = 50)	*Making Fixing Things*	3.73***		
	Exposure/Event	2.48*	R = .531 R² = .282	9.034***
Other (n = 15)	*Improve World*	2.47*	R = .551 R² = .304	6.11*
African American Male	NA			
African American Female	NA			
Hispanic Male (n = 31)	Math Skill	−4.70***		
	Opportunities	2.59*		
	Improve World	−2.36*	R = .699 R² = .489	8.92***
Hispanic Female (n = 16)	*Making Fixing Things*	2.76*		
	Math Skill	2.21*	R = .700 R² = .491	6.74**

Note. Variables positively correlated with GPA in italics.
* $p < .05$
** $p < .01$
*** $p < .001$

Appendix 5.3

Student Responses Associated With Academic Performance (GPA): What Does It Take to Be Successful in Your Engineering Program?

Group	Variable	t-Value	Multiple R/ Variance Accounted For	F-Value
All (n = 129)	Scholarship	−2.54*		
	Discipline	−1.99*	R = .266 R^2 = .071	4.83**
Male (n = 74)	Discipline	−3.07**	R = .338 R^2 = .114	9.39**
Female (n = 54)	Scholarship	−2.43*		
	Network w/ Others	*2.19*		
	Programs (MEP)	*2.11*	R = .502 R^2 = .252	5.74**
African American (n = 63)	Scholarship	−2.54*	R = .307 R^2 = .094	6.45*
Hispanic (n = 49)	Discipline	−2.42*	R = .330 R^2 =.109	5.87*
Other (n = 15)	*Study*	*5.78***		
	Discipline	−4.88***		
	Dedication/ Motivation	*3.77**		
	4-Point Learning System	*2.57*	R = .930 R^2 = .866	17.71***
African American Male (n = 32)	Study	−2.10*	R = .353 R^2 = .125	4.42*
African American Female (n = 30)	Scholarship	−2.23*	R = .382 R^2 = .146	4.97*
Hispanic Male (n = 31)	Discipline	−2.30*	R = .338 R^2 = .150	5.31*
Hispanic Female (n = 17)	*Desire*	*2.42*	R = .517 R^2 = .268	5.85*

Note. Variables positively correlated with GPA in italics.
* $p < .05$
** $p < .01$
*** $p < .001$

Appendix 5.4

Student Advice Correlated With Academic Performance (GPA): What Advice Would You Give to Other Incoming Minority Students?

Group	Variable	t-Value	Multiple R/ Variance Accounted For	F-Value
All (n = 98)	*Dedication/ Motivation*	2.24*		
	Time/ Management CW	–2.34*	R = .330 R² = .109	5.86**
Male (n = 52)	*Dedication/ Motivation*	2.39*		
	Study Group CW	–2.32*	R = .448 R² = .201	6.29**
Female (n = 45)	Time Management	–2.03*	R = .293 R² = .096	4.13*
African American (n = 36)	Faculty	–3.73***		
	Get Help CW	–2.56*		
	Work/Effort	–2.20*	R = .560 R² = .314	8.08***
Hispanic (n = 30)	*Dedication/ Motivation*	2.72**	R = .451 R² = .111	7.41**
Other (n = 10)	*Campus Involvement*	2.91*	R = .696 R² = .485	8.48*
African American Male (n = 28)	Study Group CW	–2.71*	R = .462 R² = .214	7.34*
Hispanic Male	NA			
African American Female (n = 27)	Work/Effort	–4.72***		
	Teachers CW	–4.89***		
	Schedule CW	–4.05***		
	Get Involved CW	2.95**		
	Network	2.86**	R = .861 R² = .741	12.61***
Hispanic Female (n = 13)	Network	–2.38*	R = .567 R² = .322	5.69*

Note. Variables positively correlated with GPA in italics. "CW" denotes critical word coding as opposed to thematic coding.
* $p < .05$
** $p < .01$
*** $p < .001$

Appendix 5.5

Student Program Component Evaluations Associated With Academic Performance (GPA)

Group	Variable	t-Value	Multiple R/ Variance Accounted For	F-Value
All (n = 124) Program Evaluations	Summer Bridge	−2.52*		
	Project/ Problem- Based Courses	2.35*	R = .271 R^2 = .074	4.84**
Male	NA			
Female (n = 52)	*Internship*	2.77**		
	Summer Bridge	−2.29*	R = .415 R^2 = .172	5.20**
African American	NA			
Hispanic	NA			
Other	NA			
African American Male	NA			
Hispanic Male	NA			
African American Female	NA			
Hispanic Female (n = 15)	*Internship*	2.65*	R = .578 R^2 = .334	7.02*

Note. Variables positively correlated with GPA in italics.
* $p < .05$
** $p < .01$
*** $p < .001$

6 The College Adjustment of Minorities and Non-Minorities in Engineering

Summary

Compared to non-minorities, will minorities in undergraduate engineering show the same inequities in performance and adjustment that are typical of minorities in higher education in general? To answer this question, online surveys on the undergraduate engineering experience were administered to 632 minority students (largely African American and Hispanic) and 513 non-minority students (i.e., White, Asian, and Middle Eastern). As is typical of higher education studies, non-minority students entered engineering with better standardized test scores and achieved higher grades. With test scores controlled, so that students with roughly equal abilities could be compared, there was little difference in overall GPA. However, minority students participated significantly more in academic support programs, and scored higher on all adjustment measures where there were differences. Despite these average differences, the success factors for both groups were remarkably similar: test scores; academic management skills (including metacognitive skills); good college adjustment; and participation in academic support programs. Non-minority students had greater access to internships than minorities, but not undergraduate research experiences—critical success factors for both groups. While faculty relationships were a significant success factor for non-minorities, they were not for minorities, who appear to substitute immersion in MEP programming. The MEP adjustment advantage also appears to enable minority students to perform nearly as well as non-minorities, despite poorer test scores at entry.

Introduction

Previous chapters have documented results from the first two phases of this research. Chapter 3 described a study of 26 NACME institutions that found minority and non-minority students of matched abilities were retained and graduated at the same rates. In Chapter 4, the focus group phase of the research profiled students attending 11 institutions who were described as pro-active problem solvers with science inclinations and mathematical abilities, seeking better lives and a better world through engineering. The focus group mini-survey in Chapter 5 also identified hands-on experience in engineering through project-based courses, research, and industry internships as critical in catalyzing and sustaining success in this field. This chapter describes results from the third phase of the NACME study—an online survey of program participation and adjustment of minority and non-minority students in 18 institutions.

In addition to the search for what might enhance the progress of minority students in engineering is the parallel question of how similar or different might minority engineers in college be compared to the general population of minority students in college. Further, are there differences in adjustment for minority and non-minority students, and does adjustment, differential or otherwise, have consequences for performance?

Background

Minorities in Engineering

The substantial literature on minority students in college suggests that the issue of adjustment to the college environment is a critical one, that adjustment issues are greater for minority students, and that race/ethnicity usually occasions less friendly treatment (Fries-Britt & Turner, 2002; Hurtado et al., 2006; Pascarella & Terenzini, 2005; Rovai, Gallien, & Wighting, 2005; Smith, 2009). In general, minority students exhibit poorer adjustment than non-minority students, which is said to account for their poorer performance and retention more so than differences in entering qualifications such as standardized test scores. The nature of their differential adjustment includes less positive peer interactions, restricted extra-curricular involvement, a less satisfactory social life, and in particular, impoverished interactions with faculty (Ferguson, 2003; Smith & Moore, 2002; Solórzano, Ceja, & Yosso, 2000). While the presence of covert and overt prejudice and racism colors the social interactions of minority students, experiences in the classroom are said to be more damaging because many nonminority professors and students harbor negative stereotypes about minority students' academic ability and potential that result in neglect (Jussim & Harber, 2005; Tennenbaum & Ruck, 2007). Differences have been found between the adjustment of Black and Hispanic students, such that Hispanic students may be subjected to milder forms of prejudice due to lighter skin color that makes them less visible targets (Ancis, Sedlacek, & Mohr, 2000; Oliver, Rodriguez, & Mickelson, 1985; Suarez-Balcazar, Orellana-Damacela, Portillo, Rowan, & Andrews-Guillen, 2003). Also, while Black students show far better outcomes in HBCUs compared to PWIs, Hispanic students show these differences to a lesser degree in HSIs compared to PWIs (Bridges,

Kinzie, Laird, & Kuh, 2008). This may be due to the fact that HSIs are so designated by virtue of significant Hispanic enrollment, rather than a mission to uplift Hispanic students (Gasman, 2008).

While the number of investigations of adjustment factors among minorities in science is small but growing (see Hurtado et al., 2007; Lewis, 2003), few such studies of adjustment have been conducted with minority engineering students (Brown, Morning, & Watkins, 2005; Lee, Matusovich, & Brown, 2014; Person & Fleming, 2012). The adjustment factor may be no less critical for minorities in engineering and may well be more so because of the hostile climate and resulting ethnic isolation so often described in the literature (Brown et al., 2005; Landis, 2005). Indeed, comfort in the scientific milieu has been singled out as the most important factor in the adjustment of minorities in science, a factor twice as important as positive student-faculty contact (Person & Fleming, 2012). Minority engineering students, however, live in a different world than other minority college students. They have stronger academic qualifications and must survive a more rigorous undergraduate curriculum. They are also different in that engineering schools have been encouraged by corporations and educational organizations to keep in place Minority Engineering Programs for the last four decades to facilitate the success and graduation of minorities in engineering. A predictable set of core program components cuts across the engineering institutions with students supported by NACME, which are available for all students but strongly encouraged by MEP program directors for minority students. These programs not only foster the development of academic skills, but also provide student access to support systems within engineering (Fleming, 2012a). So, will the adjustment profiles of minority engineering students resemble those of other minority students, or will they be more defined by the high-achieving environment in which they matriculate?

Non-Minorities in Engineering

The inclination toward engineering appears to be a combination of factors, including a positive commitment to engineering, enjoyment of math and science, confidence in mathematical and scientific abilities, and an impersonal personality style (Besterfield-Sacre, Moreno, Shuman, & Atman, 2001; Capretz, 2003; Godfrey & Parker, 2010). Such characteristics fit the self-efficacy model proposed by Ambrose, Lazarus, and Nair (1998), who contend that the critical factor is a perceived ability to be efficacious in the field. Furthermore, students who are successful in engineering school tend to be field independent, organized, goal directed, and with good study skills, rather than being easy-going and exploratory (Rosati, 2003; Felder, Felder, & Dietz, 2002; Zimmerman et al., 2006). Traditionally, it has been White male students who exemplified these characteristics and achieved a favored position in engineering. Further, such characteristics have been compatible with the traditional engineering curriculum and traditional methods of teaching (Felder et al., 2002). Thus, the clear majority of studies comparing minority and non-minority engineering students in college find that non-minorities fare better. In the main, non-minorities exhibit better grades, with or without better test scores; better adjustment along many dimensions; and lower dropout rates. Chen and Weko (2009) reported that while similar percentages of Whites, Asians, Blacks, and Hispanics enter STEM fields, more non-minorities graduate, making retention a crucial issue. Dropout rates in engineering are high for such qualified students, owing to the rigorous curriculum, but they are lowest for non-minorities (Araque, Roldan, & Salguero, 2009; Suresh, 2006/2007). Borrego, Padilla, Zhang, Ohland, and Anderson (2005) reported that 6-year graduation rates were higher for majority students than for minority students in engineering for seven cohorts at nine southeastern universities. According to Marra, Rodgers, and Shen (2012), majority students less often cited academic reasons for leaving engineering compared to minority students, and majority students less often cited non-academic reasons such as feeling that they did not belong in engineering. Feeling a lack of belonging in engineering was the strongest factor. Indeed, majority students do not typically share the problematic issues of belonging or comfort in the engineering milieu, since it is others (i.e., minorities and women) who must adapt to the traditional White male fraternity (Baker, Tancred, & Whitesides, 2002; Besterfield-Sacre et al., 2001; Brown et al., 2005). The question for this comparative analysis is: what differences might there be between minority and non-minority engineering students that inform an investigation of success factors?

Purpose of the Study

Four specific questions guided the investigation.

First, does greater program involvement foster greater academic success, and are there specific program components that might distinguish themselves as critical success factors? Do academic support programs serve the same function for non-minority students?

Second, what aspects of college adjustment foster greater academic success, and what aspects of adjustment are more important for minority and non-minority students?

Third, what factors appear to distinguish students matriculating in engineering schools most effective at retention and graduation of their students? Are any such factors different for minority and non-minority students?

Fourth, how do seniors, or upperclassmen, compare to freshmen or underclassmen? The value added to the undergraduate education is a concern for student development in general. Thus, freshman-senior differences in program participation and college adjustment provide a cross-sectional glimpse into the value-added issue for minorities and non-minorities.

In sum, the survey of engineering students was designed to assess student reports of program participation and adjustment issues in order to further define factors that enhance student success in terms of academic performance, the experience in the most effective engineering schools, in terms of retention or longevity in engineering programs, and how minority and non-minority students might differ in these factors.

Method

An online survey was administered via Survey Monkey to undergraduate engineering students attending 18 responding institutions, yielding a total of 1145 usable surveys. The sample, shown in Table 6.1, included 632 largely African American and Hispanic students, and 513 non-minority students of White, Asian, and Middle Eastern composition. Female students comprised about 40% of minorities, and 35% of non-minorities. Students were most likely to major in Mechanical Engineering (28.8%), followed by Computer Science (15.5%), Electrical Engineering (12.0%), Civil Engineering (10.0%), Chemical Engineering (9.9%), and Biomedical Engineering (8.1%). Other majors accounted for less than 3% each of the total. Non-minorities were significantly more likely to major in Mechanical Engineering, while Minorities were more likely to major in Civil Engineering.

The measures, which included program involvement and college adjustment, were adapted from the evaluation of nine different programs in various STEM fields, as well as studies of student development in college (Felder & Silverman, 1988; Felder et al., 2002; Fleming, 1984, 2001, 2012e; Hardy, 1974; Moos, 1979; Nasim, Roberts, Harrell, & Young, 2005; Pearson, 2005; Person & Fleming, 2012; Ramseur, 1975; Sedlacek, 1998, 2004). Measures of college adjustment were grouped into two categories: those specific to college adjustment dimensions—College Adjustment, Study Habits, Academic Adjustment, Social Adjustment, Faculty Interactions, and Comfort; and those measuring personal orientations that might influence or interact with college adjustment—Learning Styles, Community Identity, and Scientific Orientation. Throughout the report, they are collectively referred to as college adjustment measures.

Four types of analysis were conducted: comparison of mean differences on study measures between minority and non-minority students; within-group analysis of variance to determine the variables associated with GPA, effective schools, and retention to the senior years; within-group regression analysis to determine the most important variables associated with each of the dependent measures of success; and a factor analytic sorting of positive predictors of success for each group of

Table 6.1 Study Population

Ethnicity	All n = 1,145	Male 62.4% n = 714	Female 37.6% n = 431	GPA	Test Score Percentile
MINORITIES:	55.2% (632)	60.3% (381)	39.7% (251)	3.03 SD = 0.49	72.2% SD = 21.7
African American	35.6% (225)	54.6% (123)	45.3% (102)	2.99 SD = 0.51	74.5% SD = 20.9
African Caribbean, African	5.5% (35)	40.0% (14)	60.0% (21)	3.10 SD = 0.53	76.1% SD = 19.2
Hispanic	58.2% (368)	65.5% (241)	34.5% (127)	3.05 SD = 0.49	70.3% SD = 22.3
Native American	0.6% (4)	75.0% (3)	25.0% (1)	3.13 SD = 0.27	89.8% SD = 3.77
NON-MINORITIES:	44.8% (513)	64.9% (333)	35.1% (180)	3.19 SD = 0.53	84.2% SD = 16.4
White	56.7% (291)	64.9% (189)	35.1% (102)	3.23 SD = 0.50	85.8% SD = 12.5
Asian	26.9% (138)	63.0% (87)	36.9% (51)	3.19 SD = 0.56	82.7% SD = 20.4
Middle Eastern	16.4% (84)	67.9% (57)	32.1% (27)	3.09 SD = 0.56	80.8% SD = 20.2

Note. Forty-three biracial students of various compositions were deleted from the analysis. Number of students in parentheses.

students. All analyses included or provided controls for standardized test scores. The details of population and method appear in Appendix 6.1; means, standard deviations, and intercorrelations of study measures appear in Appendix 6.2; and items with subscale structure appear in Appendix 6.3.

Minority vs. Non-Minority Differences in Study Variables

Table 6.2 presents the mean differences between minority and non-minority students in the study measures. It shows that the test scores of minorities were significantly lower than those of non-minority students (72.6% vs. 84.3%). Minority grades were also lower, but the disparity and effect size was not nearly as great as for test scores (3.04 vs. 3.21). With test scores controlled, minority GPA was 3.06 vs. 3.16 for non-minorities.

With test scores controlled, minorities showed stronger Program Participation, as well as greater participation in Summer Bridge, Tutoring, Supplemental Instruction, Study Groups, and Faculty Mentoring. Non-minorities reported greater participation in Freshman Orientation Courses, and far greater Internship Involvement. Non-minority students

STUDY VARIABLES	Minority Mean n = 632	Non-Minority Mean n = 513	F with Test Score Percentile	Effect Size (Eta)
DEMOGRAPHIC VARIABLES:				
GPA	3.04 (0.02)	3.21 (0.02)	70.76***	.026
GPA, With Test Score Percentile	3.06 (0.02)	3.16 (0.02)	10.29***	.009
Test Score Percentile	72.60 (0.79)	84.29 (0.90)	94.88***	.077
RAW PROGRAM PARTICIPATION	*Minorities*	*Non-Minorities*	X^2	
Middle School STEM Programs	26.3%	23.8%	ns	na
High Schools STEM Programs	38.6%	40.0%	ns	na
Summer Bridge	25.1%	10.3%	39.5***	na
Freshman Orientation Courses	50.2%	63.2%	17.2***	na
Tutoring Programs	56.0%	40.4%	28.9***	na
Study Groups	43.5%	35.1%	8.73**	na
Supplemental Instruction	49.8%	41.1%	10.5***	na
Peer Mentoring	23.8%	19.1%	ns	na
Faculty Mentoring	29.0%	22.8%	ns	na
Faculty-Guided Research	17.5%	16.6%	ns	na
Corporate Internships	25.6%	42.5%	33.1***	na
PROGRAM PARTICIPATION SCALES:	*Minorities*	*Non-Minorities*	F/p	Effect Size (Eta)
Program Participation Scale	28.7 (0.38)	27.4 (0.43)	4.77*	.004
Middle School STEM Program Participation Scale	2.86 (0.09)	2.25 (0.09)	ns	na
High School STEM Program Participation Scale	2.86 (0.09)	2.84 (0.11)	ns	na
Summer Bridge Participation Scale	2.08 (0.01)	1.57 (0.08)	45.84***	.074
Freshman Orientation Course Participation Scale	3.11 (0.09)	3.62 (0.09)	14.54***	.013
Tutoring Participation Scale	3.48 (0.09)	2.92 (0.10)	15.27***	.013
Supplemental Instruction Participation Scale	3.30 (0.09)	2.87 (0.11)	9.07**	.008
Study Group Participation Scale	3.08 (0.09)	2.74 (0.11)	5.34*	.005
Peer Mentoring Participation Scale	2.04 (0.07)	1.83 (0.08)	ns	na
Faculty Mentoring Participation Scale	2.29 (0.08)	2.02 (0.01)	5.02*	.004
Research Participation Scale	1.82 (0.07)	1.73 (0.08)	ns	na
Internship Participation Scale	2.26 (0.09)	2.98 (0.10)	26.92***	.023

(Continued)

Table 6.2 (Continued)

COLLEGE ADJUSTMENT SCALES:	Minorities	Non-Minorities	F/p	Effect Size (Eta)
College Adjustment Scale	73.67 (0.61)	71.15 (0.67)	7.45**	.006
Comfort Scale	43.31 (0.38)	44.07 (0.34)	ns	na
Academic Adjustment Scale	59.32 (0.34)	57.22 (0.38)	16.42***	.014
Engineering Faculty Interactions Scale	119.95 (0.82)	118.04 (0.91)	ns	na
Academic Management Scale	119.14 (0.84)	116.78 (0.94)	ns	na
Study Habits Scale	66.14 (0.58)	63.51 (0.64)	8.93**	.008
Social Adjustment Scale	103.58 (0.76)	102.37 (0.85)	ns	na
Scientific Orientation Scale	68.3 (0.47)	66.5 (0.52)	6.71**	.006
Community Identity Scale	98.62 (0.57)	96.66 (0.64)	4.96*	.004
Variable	Minorities	Non-Minorities	F/p	Effect Size (Eta)
LEARNING STYLES:				
Active	6.08 (0.09)	6.23 (0.10)	ns	na
Reflective	4.92 (0.09)	4.77 (0.10)	ns	na
Sensing	6.42 (0.09)	6.60 (0.11)	ns	na
Intuitive	4.59 (0.09)	4.39 (0.11)	ns	na
Visual	7.79 (0.09)	8.06 (0.09)	4.38*	.004
Auditory	3.22 (0.09)	2.39 (0.09)	4.38*	.004
Sequential	6.45 (0.08)	6.29 (0.09)	ns	na
Global	4.55 (0.08)	4.70 (0.09)	ns	na

Note. Standard error of the mean in parenthesis. Effect size (partial *Eta* squared) given for significant results. r (GPA with Test Score Percentile) = .235***
* $p < .05$
** $p < .01$
*** $p < .001$

reported 1.7 times greater Internship Involvement than minority students (25.6% vs. 42.5%).

Differences on the college adjustment measures were all in favor of minority students. Minorities reported higher scores on College Adjustment, Academic Adjustment, Study Habits, Scientific Orientation, and Community Identity. The learning styles of minorities were more Auditory and less Visual. These differences were true overall and held after controls for differences in test scores, although the effect sizes tended to be small.

In short, where there were differences, minority engineering students reported greater program involvement and better adjustment to engineering school than non-minority students on a number of assessments. While the lower test scores and somewhat lower grades of minorities are typical of comparative studies, the indications of better adjustment among minority engineering students are atypical.

The Undergraduate Engineering Experience of Minority Students

Enhancing Academic Performance

Does the overall degree of participation in program components positively affect the academic outcomes of minority engineering students? The answer is no. However, the degree of participation in certain program components does influence students' academic outcomes. The two component scales related to GPA were Research Participation and Internship Involvement. Summer Bridge Participation was negatively related. Five adjustment scales were positively related to academic performance after a control for test scores—Academic Adjustment, Academic Management, Faculty Interactions, Comfort, and Study Habits. It seems safe to say that better college adjustment was associated with better academic performance among minority engineering students.

When the program participation scales and adjustment scales were entered into a regression equation (Table 6.3), the most important positive factors were: test scores; Academic Management; Internship Involvement; and Research Participation. Simple program involvement was a negative factor. These variables accounted for 12.6% of the variance in GPA. While program component participation and better college adjustment were generally associated with better grades, the most important of these were academic management skills, and participation in program components that provide hands-on experience in engineering—research and industry internships. Note that for minority students, test scores loaded first and were thus the most important factor in GPA.

Table 6.3 Regression Analyses of Measures of Program Participation and College Adjustment on Success Variables for Minorities (n = 632)

Dependent Variables:	Independent Variables	t	R/R²	F
GPA for Minorities				
	Test Score Percentile	5.23***		
	Academic Management Scale	4.85***		
	Internship Involvement Scale	3.84***		
	Research Participation Scale	3.51***		
	Program Involvement Subscale	−3.88***	R = .355 R² = 12.6	14.01***
Effective School Rankings for Minorities				
	Internship Involvement	6.06***		
	Freshman Orientation Course	5.218**		
	Academic Adjustment Scale	3.24***	R = .375 R² = 14.1	30.89***
Academic Status i.e., Freshman-Senior Differences for Minorities				
	Research Participation Scale	5.18***		
	Internship Involvement Scale	4.85***		
	Social Adjustment Scale	3.99***		
	Supplemental Instruction Participation Scale	3.69***		
	Study Group Participation Scale	3.19***		
	Peer Mentor Participation Scale	−4.22***		
	Engineering Faculty Scale	−3.11		
	Academic Adjustment Scale	−2.99**		
	Summer Bridge Participation Scale	−2.79**		
	Test Score Percentile	−2.67**		
	Tutoring Participation Scale	−2.32*	R = .477 R² = 22.7	16.56***

Note. Positive predictors in italics.
* p < .05
** p < .01
*** p < .001

What Distinguishes Schools Effective in Graduating Minority Engineering Students?

Is there a difference in the degree of program participation, or component participation, between the best and least successful engineering schools? We have previously defined "successful" as those programs that are best at graduating minority students; that is, in terms of the percentage of students who graduate within six years. Further, this definition applies to the percentage 6-year graduation rates that were adjusted for differences in test scores between institutions (see Chapter 3). Schools in the survey sample were ranked in this manner into the top, middle, or bottom third of the distribution.

An analysis of variance with test scores controlled indicates that students in the most successful or effective schools reported greater Program Participation, more involvement in Freshman Orientation Courses, Internships, Faculty Mentoring, and Peer Mentoring. On adjustment measures, minorities in effective engineering schools reported better Academic Adjustment, College Adjustment, more positive Faculty Interactions, greater Comfort, and better Academic Management skills.

In a regression equation (Table 6.3), the most important predictors of effective schools were Internship Involvement, Freshman Orientation Course participation, and Academic Adjustment. These three variables accounted for 14.1% of the variance in effective school rankings. If MEP program participation is a factor in facilitating the successful graduation of minorities in engineering, then the more effective schools make greater use of it. The result appears to be a more auspicious academic life, as well. These schools were perceived by their students to be more supportive, to provide more faculty contact, and more faculty mentoring, such that they were more satisfied and comfortable with their educational experience.

Identifying Freshman to Senior Differences

The average GPA of 3.04 does not change with class status after test scores were controlled. Program participation increases from freshman to senior year as seniors report significantly higher rates of involvement. Seniors report more Middle School STEM Program Participation, while freshmen reported more Summer Bridge Participation. Freshmen reported more involvement in Peer Mentoring, while seniors reported greater participation in Supplemental Instruction, Study Groups, Research, and Internships. Note that some of these reports are retrospective, as for middle and high school STEM program participation, and so may not reflect the impact of the undergraduate educational experience. Three adjustment measures showed differences between freshmen and seniors. Seniors reported greater Social Adjustment, while surprisingly, freshmen reported more positive Faculty Interactions, as well as better

Academic Adjustment. An inspection of subscales of the Faculty Interactions Scale showed that seniors were less likely to report that faculty held high expectations for them, but also reported less faculty prejudice and racism. Whatever decreases positive faculty interaction, then, does not seem related to prejudice.

In a regression equation (Table 6.3), upperclassmen were distinguished by reporting greater Social Adjustment, and participation in four MEP program components: Research; Internships; Supplemental Instruction; and Study Groups. These factors accounted for 22.7% of the variance in student classification.

Success Factors

The nine unique variables contributing positive variance in the preceding regression analyses were factor analyzed (Table 6.4). The four factors extracted were: (1) Academic and Social Adjustment; (2) Program Participation Type I (e.g., Internships); (3) Program Participation Type II (e.g., Research); and (4) Test Scores vs. Study Group Participation. This sorting suggests that a combination of good adjustment (academic and social), academic management skills, and MEP program participation contribute most to minority success in engineering.

In the foregoing regression analyses only 12.6%–22.7% of the variance could be accounted for in each of the dependent variables. To maximize the variance accounted for, each of the summary and component subscales was allowed to enter the specific regression equation along with test scores. When

Table 6.4 Factor Analysis of Success Factors for Minority Students

Factors	Factor Loadings
I: Academic/Social Adjustment	
Academic Management Scale	.850
Academic Adjustment Scale	.819
Social Adjustment Scale	.760
II: Program Participation I	
Internship Participation Scale	.730
Supplemental Instruction Scale	.672
III: Program Participation II	
Research Participation Scale	.669
Freshman Orientation Success Course Participation Scale	−.647
IV: Test Scores vs. Study Groups	
Test Score Percentile	.817
Study Group Participation Scale	−.493

this was accomplished for GPA, effective school rankings, and class status, a pool of 17 variables emerged, which then accounted for 22.8%–31.6% of the variance in the dependent variables. A factor analytic sorting of all scales and subscales produced results similar to summary scales.

Conclusions

An extensive literature on minority students in college maintains that minorities display poorer adjustment than other students, and that adjustment issues are a primary factor in their academic performance and retention. The present study of minority engineering student adjustment also suggests that aspects of college adjustment play an important role in their success. Academic management skills, satisfaction with the engineering school, and a positive adjustment to peers all contribute to better academic performance and retention; these are also keys to effective engineering school environments. Students in engineering schools more effective in graduating minority students also reported significantly more program involvement than students in other institutions, indicating that access to and immersion in MEP program components make a difference in minority student graduation rates. The most important of these program offerings are industry internships. Internship involvement was associated with all three measures of success; research involvement was associated with two measures of success. Note that seniors reported less positive interactions with faculty members than freshmen, which does suggest some faculty disenchantment even though seniors did not report more prejudice or racism. Studies of students in college, regardless of ethnicity, also report that academic management skills are important to success, perhaps more important than test scores (Sedlacek, 1998, 2004). Such skills, it is said, are the keys to negotiating the college environment successfully. Such skills also appear to be important for minority engineering students. They include effort, time management, and a meta-analytic focus on organizing engineering information.

What may distinguish minority engineering students from other minority students, apart from the rigorous curriculum, are Minority Engineering Programs with their myriad of available program components, not to mention the MEP Program Directors and staff who oversee them and encourage student participation. While program participation in general does not seem to enhance performance in this study, several program components were routinely associated with greater success. Industry internships head the list, followed by research participation, supplemental instruction, study groups, and freshman orientation success courses. While research participation was a major factor in student success, it does not appear that students at all schools, even the most effective ones, provide the same level of access to faculty guided research opportunities. Only 17% of minority engineering students indicated participation in faculty-guided research. Evaluations of highly successful programs for minority students in science and engineering have contributed a list of essential program ingredients that enhance performance and retention, including a strong demand for excellence, mentoring, research experiences, summer bridge experiences, enhanced mathematics skills, and scholarships. The list of ingredients that work is extensive, but is notable for the absence of mention of industry internships. In this study, internships were clearly the most important program component that positively influenced every aspect of "success" investigated. Research experiences were also prominent, but relatively unavailable to the clear majority of minority students. Also, note that while summer bridge and mentoring were not universal success factors in this study, they were among the program components that distinguished engineering schools most effective in graduating minority students. Finally, note that math skills—or rather, student assessments of their math competence—found no place in this set of success factors. This may be because test scores were controlled, and/or because math is the well-documented primary pathway to engineering (e.g., Pearson & Miller, 2012), such that engineering students all share strong math competence. Math, then, may not discriminate among those students who are successful; a similar argument could be made for their scientific orientation.

For all the analyses conducted, test scores were taken into account; they were either included as a variable, as in regression analyses or controlled, as in analysis of covariance. In all analyses, it was evident that test scores accounted for a significant portion of the variance in grades. However, test scores were not significantly higher in schools most effective in graduating minority students, and seniors did not have higher test scores than freshmen. It did appear, however, that test scores were associated with academic confidence.

In short, while adjustment issues play an important role in the success of minority engineering students, participation in Minority Engineering Programs plays just as critical a role in their success.

The Undergraduate Engineering Experience of Non-Minority Students

Enhancing Academic Performance

The analysis of covariance showed that for non-minority students, overall program involvement bore no relationship to GPA, nor did any of the program component scales. However, the following four adjustment measures did show significant positive relationships to GPA—Academic Adjustment, Academic Management, Faculty Interactions, and Scientific Orientation. Thus, a number of adjustment dimensions were associated with better performance among non-minority students, but not program participation. Indeed, components of program participation were typically associated with *lower* GPA for non-minority students.

In a regression equation (Table 6.5), the following three variables were the most important positive predictors of GPA: Academic Management; test scores; and Faculty Interactions. Entered variables accounted for 12.5% of the variance in GPA. Managerial academic skills were the most important factor in academic performance, more so than test scores. Such managerial skills have been associated with the ability to negotiate the academic institution successfully among minorities, and appear to have predictive power for non-minorities, as well (Sedlacek, 1998, 2004). For non-minority students, positive faculty interactions enhanced GPA, as has been found in numerous studies of college students.

What Distinguishes Schools Effective in Graduating Non-Minority Engineering Students?

Non-minority students in effective schools showed greater Program Participation, as well as greater participation in Summer Bridge and Supplemental Instruction, with the strongest effect for greater Internship Involvement. On adjustment measures, non-minority students in more effective schools reported higher scores on Social Adjustment, College Adjustment, Academic Adjustment, Comfort, Faculty Interactions, and Scientific Orientation. More effective schools, then, have more satisfied students, more engaged and comfortable in the scientific milieu. Most notably, whereas program involvement had no effect on GPA for non-minorities, it proved prominent in the factors distinguishing effective schools.

In a regression equation (Table 6.5), the four most important positive factors were—participation in Internships, better College Adjustment, more Summer Bridge Participation, and more Active Learning Styles. Negative factors, or effects least characteristic of effective schools, were—higher scores on Study Habits, Research Participation, and High School Program participation. These variables accounted for an impressive 57.8% of the variance in effective school rankings for non-minority students and were more important than test scores. In sum, non-minority students in effective schools showed greater program involvement, especially in internships, and a widespread pattern of better adjustment, especially adjustment to the engineering school environment. Note that while Internship Involvement was a significant factor for both groups of students, it was strongest for non-minorities. Indeed, the effects for non-minorities were generally stronger than for minorities with vastly more variance explained.

Identifying Freshman to Senior Differences

Among non-minority students, seniors did not report more program participation in general, but did report more participation in Research,

Table 6.5 Regression Analyses of Measures of Program Participation and College Adjustment on Success Variables for Non-Minorities (n = 513)

Dependent Variables:	Independent Variables	t	R/R^2	F
GPA for Non-Minorities				
	Academic Management Scale	4.15***		
	Test Score Percentile	4.09***		
	Engineering Faculty Scale	2.28*		
	Active Learning Styles	−2.90**		
	Tutoring Participation Scale	−2.28*		
	Freshman Orientation Course Scale	−2.22*	R = .353 R^2 = 12.5	12.04***
Effective Schools Rankings for Non-Minorities				
	Internship Involvement	22.45***		
	College Adjustment Scale	3.24**		
	Freshman Orientation Course Scale	2.58**		
	Summer Bridge Participation Scale	2.52*		
	Active Learning Styles	2.52*		
	Study Habits Scale	−4.53***		
	Research Participation Scale	−2.88**		
	High School Program Participation Scale	−2.38*	R = .760 R^2 = 57.8	75.35***
Academic Status; i.e., Freshman-Senior Differences for Non-Minorities				
	Research Participation Scale	6.77***		
	Internship Involvement Scale	3.65***		

(Continued)

Table 6.5 (Continued)

Dependent Variables:	Independent Variables	t	R/R²	F
	Study Group Participation Scale	2.71**		
	Social Adjustment Scale	2.14*		
	Engineering Faculty Scale	−4.19***		
	Freshman Orientation Course Scale	−2.69**		
	Summer Bridge Participation Scale	−2.47*		
	Test Score Percentile	−2.05*	R = .417 R² = 17.4	13.27***

Note. Positive predictors in italics.
* $p < .05$
** $p < .01$
*** $p < .001$

Internships, and Study Groups. Freshmen reported more participation in Freshman Orientation Courses and Tutoring. There were no freshman-senior differences on any of the adjustment measures.

In a regression equation (Table 6.5), senior status was most positively associated with: Research Participation; Internship Involvement; Study Group Participation; and better Social Adjustment. Freshmen reported higher scores on the Faculty Interactions Scale, Freshman Orientation Course Participation, and Summer Bridge Participation, while they also had somewhat higher test scores. The variables entered accounted for 17.4% of the variance. Critical program participation with attendant social benefits emerge as distinguishing features of retention to the senior year (if not higher GPA). More striking is the finding that less frequent or positive contact with engineering faculty among seniors is uncommon in college adjustment studies and seems inconsistent with the theoretical importance of faculty contact in past research. Most curious is that non-minority seniors reported less positive faculty interactions, but faculty interactions were an important factor in higher grades for them.

Success Factors

The 11 positive predictors of success measures were entered into a factor analysis that isolated the following three success factors, shown in Table 6.6, were: (1) College Adjustment; (2) Active Group Learning;

Table 6.6 Factor Analysis of Success Factors for Non-Minority Students

Factors	Factor Loadings
I: College Adjustment	
College Adjustment Scale	.841
Engineering Faculty Interactions Scale	.840
Social Adjustment Scale	.735
Academic Management Scale	.698
II: Active Group Learning	
Active Learning Styles	.711
Study Group Participation Scale	.659
III: Entry Level Program Participation vs. Test Scores	
Summer Bridge Participation Scale	.703
Test Score Percentile	−.631
Freshman Orientation Success Course Participation Scale	.513

Note. Research and Internship Participation scales did not load on any of the factors for non-minority students.

and (3) Entry Level Program Participation vs. High Test Scores. Note that in this sorting, adjustment and academic management skills come first and take precedence over test scores. Second, interactive or perhaps hands-on learning in groups comprise important avenues for success. Third, entry level program participation—such as in Summer Bridge and Freshman Orientation Course participation—was associated with lower test scores. Neither research nor internship participation loaded on any of the factors extracted.

Again, all scales and subscales were allowed to enter each of the regression equations in an effort to increase the variance estimates from 12.5%, 57.8%, and 17.4%. When accomplished, the resulting healthy estimates were 37.0%, 58.9%, and 58.9%, respectively.

Conclusions

The investigation of success factors for non-minority students in engineering calls attention to four main points. First, academic skill levels as in managerial skills that help students negotiate the undergraduate engineering environment are important to academic performance, more so than test scores. This kind of finding appears to cut across ethnicity. Second, while measures of good adjustment to college were unimportant as far as academic performance was concerned, adjustment measures were distinguishing features of effective engineering schools, and of retention to the senior years. Third, program involvement was also unimportant in enhancing academic performance, but was prominent in distinguishing effective schools and retention. Fourth, the results for non-minority engineering students bespeak an odd relationship to faculty. Faculty were a critical factor in enhanced academic performance, but seniors reported less positive interactions with faculty. Furthermore, positive faculty interactions were more characteristic of effective schools, but their importance was superseded (in a regression equation) by other factors. These results bear further investigation.

The Comparative Educational Experience of Minority and Non-Minority Students

In any comparison of minority and non-minority students, we expect that non-minorities might occupy a more favorable position. Certainly, this is the case in the comparative studies reviewed here, whereby non-minorities have higher test scores, better grades, higher graduation rates, and a more positive educational experience. There were also reports of better adjustment to the scientific milieu.

In this study, there were also indications that non-minority engineering students possessed a more positive profile. Their test score percentiles were 16.1% higher (72.6% vs. 84.3%). Their grades were 5.6% higher (3.04 vs. 3.21)—but not as high as expected, given the disparity in test scores. With test scores controlled, there was little difference in grades: 3.06 vs. 3.16. The one other clear advantage owing to non-minorities was their 167% greater involvement in industry internships, according to their reported participation. Since internships surface repeatedly as a significant variable in both individual and institutional success, they represent a major success factor.

However, in other respects, minority engineering students do not appear to operate at a disadvantage. In comparing participation in engineering program offerings, there were significant differences found on eight components, with minority students reporting greater involvement in six of these—the exceptions being internships and freshman orientation courses. On the assessments of adjustment, there were group differences on five measures, all of them indicating more positive adjustment among minority students to the engineering environment. On two other measures which figured prominently in individual and institutional success, academic management skills and faculty interactions, there were no differences between the two groups. In the alleged chilly climate of engineering, there were no differences in student reports of their social adjustment.

We could argue thusly: while there are systemic differences in the entering qualifications of minority engineering students that place them at a disadvantage of roughly 16 percentage points, this initial disadvantage quickly dissipates as much of the gap in academic performance decreases and their adjustment advantage increases.

Given this state of the profile, we could ask if the adjustment advantages of minority engineering students are not directly related to their greater involvement in program component offerings. Many of the program opportunities available are the direct result of existing or former Minority Engineering Programs established to facilitate the retention of under-represented students through the engineering pipeline to graduation. As such, minority students in the NACME group of institutions are overseen by program directors who encourage program participation among these students, as well as others who might benefit. It does appear that greater program participation aids the success of minority students in that it is associated with higher grades, with retention or longevity in the engineering school pipeline, and that students in schools more effective in graduating minorities report greater involvement in program activities. Although interactions with faculty were not a factor in minority student success, participation in MEP programming may provide the support that they might otherwise lack.

It also seems that when non-minority students participate in program activities, that they do so because of poor grades. Internship involvement was the only real exception to this rule. Minority students, on the other hand, appear to participate as more of a routine matter that acts to enhance grades. It should be pointed out that many of the program activities have highly interactive aspects that seem to promote academically beneficial social interaction. Other researchers have found that minority students benefit from forms of interactions that provide beneficial academic outcomes, including the "social capital" assistance from knowledgeable individuals, that can act as substitutes for faculty interaction and learning communities (Dika, Pando, & Tempest, 2016; Plett, Lane, & Peter, 2016).

Apart from average differences on study measures, the factors facilitating success in the engineering environment were remarkably similar for minority and non-minority students, with relatively minor differences. Test scores were somewhat more important to GPA for minorities than non-minorities, though they were important for both groups. This makes sense in that minority students matriculate among students who, on the average, have higher test scores, such that any test score advantage would assist academics. Managerial skills were critical for both groups, but slightly more so for non-minority students. Perhaps the most noticeable difference is that where positive faculty interactions constitute a success factor for non-minorities in academic performance, participation in the program components of internships and research were found instead for minorities. Finally, while minority students reported greater program participation, program involvement constituted a success factor for both groups.

In short, despite initial deficits in test scores, MEP programming enables minority students to make a good adjustment to the engineering school environment, better than that observed for non-minority students, with the effect that minority student performance rivals that of non-minority students. The results for this comparative study may well be different than for minorities in other engineering schools, simply because these institutions were selected for the NACME group due to their strong programming for minority students.

On a methodological note, the difference between adjustment measures and skill measures should be clarified, specifically the difference between academic adjustment and academic management. Adjustment measures generally refer to the student's feelings of satisfaction, while other measures refer to reported estimates of skill levels—in other words, feelings rather than competencies. Academic adjustment refers to satisfaction with aspects of the college experience, such as happiness with the choice of college, feelings of confidence, and compatibility with the course of study. Academic management refers to skills that are managerial in nature and enable students to negotiate the educational experiences successfully, such as the meta-analytic inclination to organize information into graphs and charts, or to use the schedule as a problem-solving tool, or to take a pro-active approach to difficulties such as problematic instructors, or racism. Both academic adjustment and academic skills played a prominent role in predicting success for minority and non-minority students, but adjustment measures play a stronger role for minorities in engineering. Also, the study was unable to account for as much variance in effective school rankings for minorities as for non-minorities—16.8% vs. 58.9%. For other success variables, a respectable one-third of the variance was achieved. While the predictive independent variables were theoretically informative, the low variance estimate suggests that something important is missing from the variable mix. The estimate was not improved by the consideration of social class, level of financial support, or the number of hours worked. Thus, further research is needed to search for more factors that might explain why effective schools for minorities are so.

References

Ambrose, S., Lazarus, B., & Nair, I. (1998). No universal constants: Journeys of women in engineering and computer science. *Journal of Engineering Education*, 87(4), 363–368.

Ancis, J. R., Sedlacek, W. E., & Mohr, J. J. (2000). Student perceptions of campus climate by race. *Journal of Counseling and Development*, 78(2), 180–186.

Araque, F., Roldan, C., & Salguero, A. (2009). Factors influencing university dropout rates. *Computers and Education*, 53(3), 563–574.

Astin, A. W. (1999). Student involvement: A developmental theory for higher education. *Journal of College Student Development*, 40(5), 518–529.

Baker, S., Tancred, P., & Whitesides, S. (2002). Gender and graduate school: Engineering students confront life after the B.Eng. *Journal of Engineering Education*, 91(1), 41–48.

Besterfield-Sacre, M., Moreno, M., Shuman, L. J., & Atman, C. J. (2001). Gender and ethnicity differences in freshmen engineering student attitudes: A cross-institutional study. *Journal of Engineering Education, 90*(4), 477–490.

Borrego, M. J., Padilla, M. A., Zhang, G., Ohland, M. W., & Anderson, T. J. (2005). *Graduation rates, grade-point average, and changes of major of female and minority students entering engineering.* Proceedings, ASEE/IEEE Frontiers in Education Conference, October 19–22, Indianapolis, IN.

Bridges, B. K., Kinzie, J., Laird, T. F. N., & Kuh, G. D. (2008). Student engagement and student success at historically Black and Hispanic-serving institutions. In M. Gasman, B. Baez, & C. S. V. Turner (Eds.), *Understanding minority-serving institutions* (pp. 217–236). Albany, NY: State University of New York Press.

Brown, A. R., Morning, C., & Watkins, C. (2005). Influence of African American engineering student perceptions of campus climate on graduation rates. *Journal of Engineering Education, 94*(4), 263–271.

Capretz, L. F. (2003). Personality types in software engineering. *International Journal of Human-Computer Studies, 58*(2), 207–214.

Chen, X., & Weko, T. (2009). *Students who study science, technology, engineering and math (STEM) in postsecondary education.* Washington, DC: U.S. Department of Education.

Dika, S. L., Pando, M. A., & Tempest, B. (2016). *Investigating the role of interaction, attitudes, and intentions for enrollment and persistence in engineering among underrepresented minority students.* Proceedings of the 2016 Annual Conference of the American Society of Engineering Education (ASEE), New Orleans, LA. Retrieved from https://peer.asee.org/17069.

Felder, R. M., Felder, G. N., & Dietz, E. J. (2002). The effects of personality type on engineering student performance and attitudes. *Journal of Engineering Education, 91*(1), 3–17.

Felder, R. M., & Silverman, L. K. (1988). Learning and teaching styles in engineering education. *Engineering Education, 78*(7), 674–681.

Ferguson, R. F. (2003). Teachers' perceptions and expectations and the black-white test score gap. *Urban Education, 38*(4), 460–507.

Fleming, J. (1984). *Blacks in college.* San Francisco: Jossey-Bass.

Fleming, J. (2001). The impact of an historically black college on African American students: The case of Lemoyne-Owen College. *Urban Education, 36*(5), 587–610.

Fleming, J. (2002). Identity and achievement: Black ideology and the SAT in African American college students. In W. R. Allen, M. B. Spencer, & C. O'Conner (Eds.), *African American education: Race, community, inequality and achievement* (pp. 77–92). New York: Elsevier Science.

Fleming, J. (2004). The significance of historically black colleges for high achievers: Correlates of standardized test scores in African American students. In M. C. Brown II & K. Freeman (Eds.), *Black colleges: New perspectives on policy and practice* (pp. 29–52). Westport, CT: Praeger.

Fleming, J. (2012a). Retaining students in engineering at the City College of New York: How a successful program works. In J. Fleming, *Enhancing the academic performance and retention of minorities: What we can learn from program evaluation* (pp. 74–85). San Francisco: Jossey-Bass.

Fleming, J. (2012b). When underprepared students stay in college: The Fast Track Program at Texas Southern University. In J. Fleming, *Enhancing the academic performance and retention of minorities: What we can learn from program evaluation* (pp. 184–200). San Francisco: Jossey-Bass.

Fleming, J. (2012c). Problem-solving instruction: How much is enough to increase verbal and analytical skills in minority students? In J. Fleming, *Enhancing the academic performance and retention of minorities: What we can learn from program evaluation* (pp. 201–218). San Francisco: Jossey-Bass.

Fleming, J. (2012d). What successful students in science know about learning: Gateway to Higher Education. In J. Fleming, *Enhancing the academic performance and retention of minorities: What we can learn from program evaluation* (pp. 51–61). San Francisco: Jossey-Bass.

Fleming, J. (2012e). *Enhancing the academic performance and retention of minorities: What we can learn from program evaluation.* San Francisco: Jossey-Bass.

Fleming, J. (2017). *City College of the City University of New York Louis Stokes Alliance for Minority Participation: Evaluation report.* Unpublished manuscript, New York NY: City College.

Fries-Britt, S., & Turner, B. (2002). Uneven stories: The experiences of successful Black collegians at a historically Black and a traditionally White campus. *Review of Higher Education, 25*(3), 315–330.

Gasman, M. (2008). Minority-serving institutions: A historical backdrop. In M. Gasman, B. Baez, & C. S. V. Turner (Eds.), *Understanding minority-serving institutions* (pp. 18–27). Albany, NY: State University of New York Press.

Godfrey, E., & Parker, L. (2010). Mapping the cultural landscape in engineering education. *Journal of Engineering Education, 99*(1), 5–22.

Hardy, F. R. (1974). Social origins of American scientists and scholars. *Science, 185*(4150), 497–586.

Hurtado, S., Cerna, O. S., Chang, J. C., Saenz, V. B., Lopez, L. R., Mosqueda, C., et al. (2006). *Aspiring scientists: Characteristics of college freshmen interested in the biomedical and behavioral sciences.* Los Angeles: Higher Education Research Institute.

Hurtado, S., Han, J. C., Saenz, V. B., Espinosa, L. L., Cabrera, N. L., & Cerna, O. S. (2007). Predicting transition and adjustment to college: Biomedical and behavioral science aspirants' and minority students' first year of college. *Research in Higher Education, 48*(7), 841–887.

Jussim, L., & Harber, K. D. (2005). Teacher expectations and self-fulfilling prophecies: Known and unknowns, resolved and unresolved controversies. *Personality and Social Psychology Review, 9*(2), 131–155.

Kolb, D. A. (1984). *Experiential learning: Experience as the source of learning and development.* Englewood Cliffs, NJ: Prentice-Hall.

Landis, R. (2005). *Retention by design.* New York: National Action Council for Minorities in Engineering.

Lee, W. C., Matusovich, H. M., & Brown, P. R. (2014). Measuring underrepresented student perceptions of inclusions within engineering departments and universities. *International Journal of Engineering Education, 30*(1), 150–165.

Lewis, B. F. (2003). A critique of the literature on the underrepresentation of African Americans in science: Directions for future research. *Journal of Women and Minorities in Science and Engineering, 9*(3/4), 361–373.

Marra, R. M., Rodgers, K. A., & Shen, D. (2012). Leaving engineering: A multi-year single institution study. *Journal of Engineering Education, 101*(1), 6–27.

Moos, R. H. (1979). *Evaluating educational environments: Procedures, measures, finding and policy implications.* San Francisco: Jossey-Bass.

Nasim, A., Roberts, A., Harrell, J. P., & Young, H. (2005). Non-cognitive predictors of academic achievement for African Americans across cultural contexts. *Journal of Negro Education, 74*(4), 344–348.

Oliver, M. L., Rodriguez, C. J., & Mickelson, R. A. (1985). Brown and black in white: The social adjustment and academic performance of Chicano and Black students in a predominately White university. *The Urban Review, 17*(1), 3–24.

Pascarella, E. T., & Terenzini, P. T. (2005). *How college affects students.* San Francisco: Jossey-Bass.

Pearson, W. (2005). *Beyond small numbers: Voices of African American PhD chemists.* New York: Elsevier Science.

Pearson, W., & Miller, J. D. (2012). Pathways to an engineering career. *Peabody Journal of Education, 87*(1), 46–61.

Person, D. R., & Fleming, J. (2012). Who will do math, science, engineering and technology? Academic achievement among minority students in seventeen institutions. In J. Fleming, *Enhancing the retention and academic performance of minorities: What we can learn from program evaluation* (pp. 120–134). San Francisco: Jossey-Bass.

Plett, M., Lane, A., & Peter, D. M. (2016). *Understanding diverse and atypical engineering students: Lessons learned from community college transfer scholarship recipients.* Proceedings of the 2016 Annual Conference of the American Association of Engineering Education, New Orleans, LA. Retrieved from https://peer.asee.org/14483.

Ramseur, H. (1975). *Continuity and change in black identity: A study of black students at an interracial college.* Unpublished doctoral dissertation, Harvard University.

Rosati, P. (2003). *Student performance in chosen engineering discipline related to personality type.* Unpublished manuscript, Department of Civil Engineering, University of Western Ontario. Retrieved from www.siu.edu/~coalctr/paper301.htm.

Rovai, A. P., Gallien, L. B., & Wighting, M. J. (2005). Cultural and interpersonal factors affecting African American academic performance in higher education. *Journal of Negro Education, 74*(4), 359–370.

Sedlacek, W. E. (1998). Admissions in higher education: Measuring cognitive and non-cognitive variables. In D. J. Wilds & R. Wilson (Eds.), *Minorities in higher education 1997–98* (pp. 47–68). Washington, DC: American Council on Education.

Sedlacek, W. E. (2004). *Beyond the big test: Noncognitive assessment in higher education.* San Francisco: Jossey-Bass.

Smith, S. S., & Moore, M. R. (2002). Expectations of campus racial climate and social adjustment among African American college students. In W. R. Allen, M. B. Spencer, & C. O'Connor (Eds.), *African American education: Race community, inequality and achievement* (pp. 93–118). Burlington, MA: Elsevier Science.

Smith, W. A. (2009). Campuswide climate: Implications for African American students. In L. C. Tillman (Ed.), *The SAGE handbook of African American education* (pp. 297–309). Thousand Oaks, CA: Sage.

Solórzano, D., Ceja, M., & Yosso, T. (2000). Critical race theory, racial micro-aggressions, and campus racial climate: The experiences of African American college students. *Journal of Negro Education, 69*(1/2), 60–73.

Suarez-Balcazar, Y., Orellana-Damacela, L., Portillo, N., Rowan, J. M., & Andrews-Guillen, C. (2003). Experiences of differential treatment among college students of color. *The Journal of Higher Education, 74*(4), 428–444.

Suresh, R. (2006/2007). The relationship between barrier courses and persistence in engineering. *Journal of College Student Retention, 8*(2), 215–239.

Tennenbaum, H. R., & Ruck, M. D. (2007). Are teacher's expectations different for racial minority than for European American students? A meta-analysis. *Journal of Educational Psychology, 99*(2), 253–273.

Tang, J. (2000). *Doing engineering: The career attainment and mobility of Caucasian, Black and Asian American engineers.* Lanham, MD: Row and Littlefield.

Zimmerman, A. P., Johnson, R. G., Hoover, T. S., Hilton, J. W., Heinemann, P. H., & Buckmaster, D. R. (2006). Comparison of personality types and learning styles of engineering students, agricultural systems management students, and faculty in an agricultural and biological engineering department. *Transactions of the American Society of Agricultural Biological Engineers (ASABE), 49*(1), 311–317.

Appendix 6.1
Methodological Details

Student Population

Table 6.1 describes the student population. It indicates that there were 632 minority students who comprised 55.2% of the total sample of undergraduate engineering students. The ethnic composition included 225 (or 35.6%) African American, 35 (or 5.5%) other students of African descent, 368 (or 58.2%) Hispanic students, and 4 (or 0.06%) Native American. By Gender, 381 (or 60.3%) were males, and 251 (or 39.7%) were female. Note that 43 biracial students of various compositions were omitted from the investigation of ethnic differences, and were included only in whole-sample analyses.

Non-minorities in this investigation included a majority of White students, but substantial numbers of non-White students. This group of 513 non-minority students was 56.7% White, 26.9% Asian, and 16.4% East Indian and Middle Eastern. Thirty-five percent were female (see Table 6.1). Some authors may argue with the composition of this non-minority group. For example, Tang (2000) would classify Asian students as disadvantaged, since she presents evidence that among engineers, Blacks do not fare significantly worse than Asians, and Asians do not fare significantly better than Blacks. While this point is well taken, Tang's arguments refer to experiences in the workplace rather than to issues of entry into engineering, undergraduate experiences, or retention. While this study was designed to compare the three disadvantaged groups with all other students, there remain opportunities for disaggregating the non-minority population, as in Chapter 8 on gender.

Measures

In addition to a background questionnaire, the 11 measures chosen for the survey fell roughly into three categories: measures of MEP program participation; measures of adjustment to the undergraduate engineering environment; and measures personal orientation that may interact with college adjustment. Student Consent included consent to retrieve official overall GPA and test scores from their institutions. Means, standard deviations, and intercorrelations of study measures appears in Appendix 6.2. Items of the survey measures along with the subscale structures are presented in Appendix 6.3.

1. *The MEP Program Participation Assessment* (Fleming, 2012a). From the preceding investigation of program components and offerings in the network of NACME programs, a generic list of components was developed. However, this aspect of the analysis was concerned with more than the existing program composition. It was concerned with whether students actually participate in the components, the degree to which they participate, and the effects of greater participation. This approach was suggested by Astin's (1999) developmental theory of student involvement, which posits that it is the degree of physical and psychological investment that the student devotes to the academic experience that determines whether the desired learning outcome is achieved. Previous evaluations have had difficulty in attempting to assess the effect of any one program activity, because not all students participate in all activities, and they do not participate to the same degree. Therefore, the present strategy, adapted from an evaluation of Program for Retention of Engineering Students at the City College of New York (Fleming, 2012a), was to design a questionnaire that includes participation questions on all program activities, so that an overall assessment of MEP participation could be made. This assessment was thus not dependent on any given pattern of program activity. A resulting MEP Program Participation Scale was constructed to measure Participation, Satisfaction, and Performance aspects of each program component separately and in sum for the 11 program components, including a retrospective assessment of participation in middle and high school STEM programs. Thus, for each program component, students were asked whether or not they participated (yes or no), their degree of satisfaction with the program component, and the extent to which they felt the component helped to improve their performance (on a 6-point Likert-type scale). Responses to rated questions were truncated to low or high categories to permit summation into a total scale. Thus, for each component, total scores ranged from 0 to 3, and total scores could range from 0 to 33. Subscales were developed for participation, satisfaction and performance across components separately. Further, a mini-scale was created for each program component separately. The whole scale estimate of internal consistency was .822; estimates for similar previous scales were .822 (Fleming, 2012a), and .824 (Fleming, 2017). Item-scale correlations ranged from .236 to .453. Among undergraduate engineering students in their MEP program at City College, this scale was unrelated to grades, but positively related to better study habits and more

positive faculty interactions. Among undergraduate STEM students in City College, the Program Participation Scale was associated with lower grades, but higher academic adjustment, faculty interactions, and scientific orientation (Fleming, 2017).

Measures of College Adjustment

Measures of college adjustment were grouped into two categories: those specific to college adjustment dimensions; and those measuring personal orientations that might influence or interact with college adjustment. Throughout the report, they are collectively referred to as college adjustment measures. The adjustment areas, suggested by previous research, were:

2. *College Adjustment* (Fleming, 2001). Suggested by the Moos (1979) assessments of college climate, 16 items composed of four subscales of four items each assessed degree of agreement on a 6-point Likert-type scale, with five items measuring perceived degree of Support, Challenge, Personal Development, and Intellectual Development. All items were stated positively. The estimate of internal consistency (i.e., alpha) for the whole scale was .946 for 1,186 students. Subscale estimates were .811, .796, .848, and .837, respectively. Other reports of whole internal consistency include .948 (Fleming, 2001), and .922 (Fleming, 2012b). Item-scale correlations ranged from .591–.811. The College Adjustment Scale has been associated with better adjustment among seniors, especially males, compared to freshman in an HBCU (Fleming, 2001), and was positively associated with 1-year retention among at-risk students at an urban HBCU, while the Challenge subscale produced the highest scores and the best prediction of 1-year retention (Fleming, 2012b).

3. *Comfort Scale* (Person & Fleming, 2012). Ten items assessed degree of comfort with the college environment, math and science faculty, and fellow students. Whole scale alpha = .863. Item-scale correlations ranged from .413–.650. The Comfort Scale was positively associated with academic performance among STEM students in a range of colleges, and was twice as important as faculty interactions.

4. *Academic Adjustment* (Fleming, 2001, 2012b). Sixteen items assess the degree of perceived adjustment to the academic experience of college. Four subscales assessed satisfaction with College, Academic Effort, Academic Confidence, and Lack of Academic Fit. Forty-three percent (7/16) of the items were stated negatively to avoid response set tendencies; these items were indicated by an R as they were reversed before scaling. Note that reversed items generally yielded lower estimates of internal consistency. Whole scale alpha

was .825; subscale estimates were .733, .817, .506, and .556, respectively. Other whole scale alpha estimates reported include .672 (Fleming, 2012b), and .838 (Fleming, 2017). Item-scale correlations ranged from .259–.580. The Academic Adjustment Scale produced low scores among at-risk students, particularly the Academic Effort subscale (Fleming, 2012b). Among STEM students, this scale was associated with greater participation in undergraduate research (Fleming, 2017).

5. *The Engineering Faculty Interactions Scale* (Fleming, 2012a) was adapted from Fleming (2001) to refer specifically to engineering faculty and consisted of 28 items assessing the perceived quality of faculty relationships. Six subscales assessed perceived Faculty Performance, Faculty Expectations of Students, Degree of Positive Faculty-Student Interaction, Satisfaction with Faculty, Faculty Mentoring of Students, Faculty Neglect of Students, and Experience of Racism. Thirty-nine percent (11/28) of the items were stated negatively to avoid response set tendencies; these items were indicated by an R as they were reversed before scaling. Internal consistency (i.e., alpha) for the whole scale was .891, and subscale estimates were .654, .713, .690, .693, .578, .744, and .558, respectively. Other whole scale alpha estimates reported include .907 (Fleming, 2001), .846 (Fleming, 2012a), .870 (Fleming 2012b), and .909 (Fleming, 2017). Item-scale correlations ranged from .207–.682. This scale has been associated with higher scores among seniors compared to freshmen in an HBCU (Fleming, 2001), with better study habits among engineering students (Fleming, 2012a), higher scores among at-risk students matriculating in an intensive academic support program (Fleming, 2012b), and greater involvement in undergraduate research among STEM students (Fleming, 2017).

6. *Study Habits* (adapted from Fleming, 2012a). Twenty items assessed perceived approaches to studying. Five subscales assessed the degree of agreement with items measuring Time Spent Studying, Structured Study Habits, Meta-Organization of Studying, Managing Interference with Study, and Study with Others. Twenty-one percent (5/24) of the items were stated negatively. Whole scale alpha was .835; subscale estimates were .632, .709, .647, .632, and .801, respectively. Other reports of internal consistency include .871 (Fleming, 2001), .853 (Fleming, 2012a), .878 (Fleming, 2012b), and .790 (Fleming, 2012c). Item-scale correlations ranged from .240–.591. The Study Habits Scale has been associated with higher scores among seniors, compared to freshmen, in an HBCU (Fleming, 2001), better faculty relationships (Fleming, 2012a), low scores among at-risk students (Fleming, 2012b), and increased scores with student-sensitive instruction in academic management (Fleming, 2012c).

7. *Academic Management* (Fleming, 2001, 2012a). Similar in theory to the work of Sedlacek (1998, 2004; see also Nasim et al., 2005), it was specifically suggested by the pro-active managerial skills found to distinguish successful STEM students from others in the Gateway to Higher Education Program (Fleming, 2012d). Twenty-eight items assessed perceived managerial approach to college, with seven sub-scales assessing Time Management, Schedule Management, Managerial Personality/Self Concept, Strategic Planning, Stress Management, Teacher Management, and Pro-Active Problem-Solving. Twenty-six percent (9/35) of the items were stated negatively to minimize response set tendencies. Whole scale alpha was .906. subscale estimates were .709, .713, .718, .580, .682, .706, and .561, respectively. Other whole scale alpha estimates reported include .809 (Fleming, 2001), .866 (Fleming, 2012b), .833 (Fleming, 2012c), and .913 (Fleming, 2017). Item-scale correlations ranged from .281–.706. This scale has been associated with better GPA (Fleming, 2001), better study habits (2012a), prediction of first-year GPA (Fleming, 2012b), increased scores with student-sensitive instruction in problem solving (Fleming, 2012c), and increased scores with greater participation in undergraduate research (Fleming, 2017).

8. *Social Adjustment Scale* (Fleming, 1984, 2001). This scale is composed of 24 items assessing quality of relationships on campus. Six subscales assessed degree of Adjustment to People, Adjustment to Friends, Satisfaction with Dating, Sense of Belonging, Leadership, and Social Management. Twenty-five percent (6/24) of the items were stated negatively to reduce response set tendencies. Whole scale alpha = .850; subscale estimates are .756, .677, .625, .771, .650, and .600, respectively. Other whole scale alpha estimates reported include .776 (Fleming, 2001), .867 (Fleming 2012b), and .831 (Fleming, 2017). Item-scale correlations ranged from .160–.706. Social Adjustment scores have been associated with *lower* grades (Fleming, 2001), low scores in an urban university (Fleming, 2012b), and greater study group participation, while the Leadership subscale was associated with greater academic program participation (Fleming, 2017).

Measures of Personal Orientation Scales

9. *Scientific Orientation.* Based on previous studies of the personal leanings of scientists and engineers (Hardy, 1974; Pearson, 2005), as well as this study's focus group findings of minority engineering student profiles, 20 items assessed Early Exposure to Math and Science, Math Competence, Problem-Solver, Knowable Universe (belief in natural laws), and Improve the World. Fifteen percent (3/20) of the items were stated negatively. Whole scale alpha was .861; subscale

estimates were .604, .797, .683, .885, and .583, respectively. Item-scale correlations ranged from .323–.587. Fleming (2017) reported a whole scale alpha of .864; this study showed that STEM undergraduates who participated in undergraduate research programs reported higher Scientific Orientation scores, and the greater the involvement in undergraduate research, the higher the scores.

10. *Community Identity.* Based on Ramseur's Black Ideology Scale (Ramseur, 1975; Fleming, 1984, 2002), a 25-item scale was adapted for use with diverse ethnic groups which assesses aspects of an individual vs. group identity. Five subscales assessed belief in Community Organization, Involvement/Agency in world issues, Community Identity Integration (feelings of social integration in the entire community), Ethnic Transcendence (feeling that people of different groups are basically the same), and Assertiveness on Behalf of Others (through fighting racism and building communities). Whole scale alpha was .851; subscale estimates were .305, 631, .744, .672, and .738, respectively. Other whole scale estimates were .842 (Fleming, 1984). Item-scale correlations ranged from .110–.635. This scale has been associated with increasing scores from freshman to senior year among Black college students (Fleming, 1984), low Black (community) identity among high achieving Black males but not females (Fleming, 2002), and low Black (community) identity among high achievers in White colleges, but not Black colleges (Fleming, 2004).

11. *Learning Styles* (Felder & Silverman, 1988; Felder et al., 2002). These have been the subject of much research aimed at understanding the personalities that best thrive in engineering, as well as how to change the curriculum to accommodate a wider array of learning styles. Adapted from Kolb's (1984) Learning Styles, the eight-pronged measure was used to assess Active, Reflective, Visual, Auditory, Intuitive, Sensing, Sequential, and Global learning styles. Note that inductive and deductive dimensions were eliminated from the inventory by Felder and Silverman (1988).

Background measures included gender, classification, SES measured through a combination of father's and mother's education, and level of scholarship support as none, partial or full.

Procedure

After finalizing the survey content in Survey Monkey, NACME program liaisons were sent a Student Information Letter describing the Survey of the Undergraduate Engineering Experience with a link to the survey for that specific institution. Program Directors then forwarded the letter to all their undergraduates, minority as well as non-minority. Students interested in receiving $25 for their effort logged in to Survey Monkey to complete it. After the first semester of the offering, there were insufficient numbers of minority students participating, so that the offering was extended to a second semester, but open only to under-represented minorities of Hispanic, African, or Native American descent.

Data Analysis

The data analysis was designed to answer three research questions: (1) What are the critical factors in enhancing academic performance? (2) What are the critical factors associated with schools more effective in graduating minority as well as non-minority students? (3) What are the critical factors associated with retention or longevity in the engineering program, or value-added? For each of these questions, the analysis determined first-order relationships with the dependent measures in question; that is, GPA, 6-year graduation rankings (for minority students as well as for non-minority students separately); and student classification, freshmen (underclassmen) and seniors (upperclassmen). Then the most critical variables were determined from a multiple regression equation. Finally, variables positively associated with the dependent "success" variables were entered into a factor analysis in order to sort the significant variables into common groupings. When necessary to account for sufficient amounts of variance, subscales were allowed to enter the regression analyses.

Appendix 6.2
Means, Standard Deviations, and Intercorrelations of Study Measures (n = 1,186)

	Program Participation Scale	College Adjustment Scale	Comfort Scale	Academic Adjustment Scale	Faculty Interactions Scale	Study Habits Scale	Academic Management Scale	Social Adjustment Scale	Community Identity Scale	Scientific Orientation Scale	GPA	Test Score Percentile
	28.1 (9.52)	72.6 (14.9)	43.7 (8.49)	58.3 (8.43)	119.1 (20.2)	65.0 (14.4)	118.1 (20.8)	103.1 (18.8)	97.7 (14.2)	67.6 (11.5)	3.10 (0.52)	77.8 (20.2)
Program Participation Scale	1.00	.308***	.318***	.229***	.277***	.301***	.267***	.331***	.212***	.183***	-.028	-.070*
College Adjustment Scale		1.00	.623***	.677***	.682***	.294***	.474***	.477***	.284***	.361***	.092**	-.069*
Comfort Scale			1.00	.563***	.580***	.350***	.539***	.540***	.347***	.381***	.091**	-.068*
Academic Adjustment Scale				1.00	.620***	.335***	.582***	.495***	.348***	.437***	.218***	-.063*
Faculty Interactions Scale					1.00	.210***	.481***	.501***	.326***	.372***	.151***	-.009
Study Habits Scale						1.00	.563***	.320***	.366***	.328**	.014	-.175***
Academic Management Scale							1.00	.530***	.462***	.532***	.191***	-.034

	Social Adjustment Scale	Community Identity Scale	Scientific Orientation Scale	GPA	Test Score Percentile
Social Adjustment Scale	1.00	.397***	.365***	.110***	-.037
Community Identity Scale		1.00	.471***	.023	-.011
Scientific Orientation Scale			1.00	.069*	-.003
GPA				1.00	.235***
Test Score Percentile					1.00

Appendix 6.3
Survey Instruments

Contents

Consent Form

Consent Form

Dear Student:

You are invited to participate in a study of success factors for students in engineering funded by the **National Science Foundation**. We are interested in what has led you to an interest in engineering and your academic experiences so far. By doing so we hope to help colleges and universities improve the student experience and graduate more engineers. Studies like this are instrumental in improving the academic support programming and instruction in engineering schools across the country.

We are asking you to:

(1) complete an online questionnaire that should take about 45–50 minutes of your time
(2) allow your academic performance data to be linked to your questionnaire responses

There are virtually <u>no risks to you in participating in this study</u> and you will be helping to make the engineering experience better for future students such as yourself.

Every effort will be made to protect your privacy by deleting your name and identifying information that would be seen by anyone other than the Principal Investigator.

For this study you will be paid $25.00 for filling out the online questionnaire.

You are free to withdraw your participation at any time. If you have questions or desire further information please contact me:

Dr. Jacqueline Fleming
JacquelineFleming@yahoo.com

I understand the nature of my participation in the "Study of the Engineering Experience" and agree to participate

_____   _____

Student Name   Date

Please fill out the following questions as completely and thoughtfully as you can. The questions concern your adjustment and satisfaction with the engineering program. Your honest answers will help the project directors design a better program experience for all parties concerned. We appreciate your cooperation.

I PROGRAM PARTICIPATION SCALE

Instructions:

The following statements refer to your participation in and satisfaction with the engineering program. Please answer the questions as truthfully as you can. Each question has three parts. If you answer NO to part A, then you may skip parts B and C. For parts B and C, please indicate a number from 1 to 6 which best fits your assessment, where 1 is not at all satisfied or no improvement and 6 is very satisfied or great improvement.

1A. Did you participate in a *special* math and science program while you were in *middle school*? _____No _____ Yes
1B. On the whole, were you satisfied with the *special* math and science middle school program?
1C. On the whole, did your middle school program help improve your math or science performance?
2A. Did you participate in a *special* math and science program while you were in *high school*? _____No _____ Yes
2B. On the whole, were you satisfied with the *special* math and science high school program?
2C. On the whole, did your high school program help improve your math or science performance?

3A.   Did you participate in a *summer bridge* program before entering your engineering program? _____No _____ Yes

3B.   On the whole, were you satisfied with the *summer bridge* program?

3C.   On the whole, did the *summer bridge* program help improve your performance in your engineering program?

4A.   Did you participate in a *Freshman Orientation course* specifically designed for engineering students? _____No _____ Yes

4B.   On the whole, were you satisfied with the *Freshman Orientation* course?

4C.   On the whole, did the *Freshman Orientation* course help improve your performance in your engineering program.

5A.   In your engineering program, have you participated in *Supplemental Instruction* review sessions outside of class conducted by peer leaders in connection with one or more of your courses? _____No _____Yes

5B.   On the whole, were you satisfied with the *Supplemental Instruction*?

5C.   On the whole, did the *Supplemental Instruction* help improve your performance in your engineering Program?

6A.   On the average, how many *tutoring* sessions do you attend per week? __none ___1 to 2 ___3 or more per week

6B.   Are you satisfied with the tutoring services?

6C.   Has participation in the *tutoring* program services improved your performance in math and science?

7A.   Do you participate in a regular study group? _____No _____ Yes

7B.   Are you satisfied with your *study group*?

7C.   Has participation in your *study group* improved your performance in your engineering program?

8A.   Do you have a *peer (student) mentor*, in conjunction with a peer mentoring program in your engineering school? _____No _____ Yes

8B.   Are you satisfied with your *peer mentor*?

8C.   Has having a *peer mentor* improved your performance in your engineering program?

9A.   Do you have a *faculty mentor*, in conjunction with a faculty mentoring program in your engineering school? _____No _____ Yes

9B.   Are you satisfied with your *faculty mentor*?

9C.   Has having a *faculty mentor* improved your performance in your engineering program?

10A. Are you working with a professor on a *research project* in your engineering school? ___No ___Yes

10B. Are you satisfied with your *research project*?

10C. Has participation in a faculty *research project* improved your performance in your engineering program?

11A. Have you participated in an *industry internship* in connection with your engineering school? _____No _____ Yes

11B. Are you satisfied with your *industry internship*?

11C. Has participation in an *industry internship* improved your performance in your engineering program?

II. COLLEGE ADJUSTMENT QUESTIONNAIRE

Instructions:

The following statements refer to your experience in your engineering program. Please answer the questions as truthfully as you can. Please indicate a number which best fits your experience, where "1" is not at all true of you and "6" is very true of you. If the question is not relevant, please circle a "1." Please use ANY number that best describes your opinion.

SUPPORT

1. The engineering faculty takes pride in its students.
5. In this engineering program, I feel as if I am not alone.
9. The community of this engineering program has given me guidance and assistance.
13. The engineering faculty here is exceptionally devoted.

CHALLENGE

2. I have been tested and challenged in this engineering program.
6. This engineering program has inspired me to give my best effort.
10. In this engineering program, I am expected to develop leadership ability.
14. In this engineering program, I have received a quality experience.

PERSONAL DEVELOPMENT

3. I have grown and evolved in this engineering program.
7. Dreams I had before coming to this engineering program are now close to being realized.
11. My talent has been greatly developed in this engineering program.
15. I am a more complete person than when I arrived at this engineering program.

INTELLECTUAL DEVELOPMENT

4. This engineering program has encouraged my intellectual development.
8. In this engineering program, I have become more competent.
12. In this engineering program, I have learned to argue effectively.
16. Here, I have become a critical thinker.

III. COMFORT LEVEL

1. What is your comfort level in Department functions?
2. What is your comfort level in meetings and clubs?
3. What is your comfort level in engineering classes?
4. What is your comfort level in other classes?
5. What is your comfort level with other engineering majors?
6. What is your comfort level in study groups?
7. What is your comfort level with students of other races?
8. What is your comfort level with the financial aid office?
9. What is your comfort level with the career services offices?
10. What is your comfort level when asking for academic assistance?

IV. ACADEMIC ADJUSTMENT QUESTIONNAIRE

I. COLLEGE SATISFACTION

1R. I am extremely unhappy with my college choice of an engineering program.
5. I have extremely positive feelings about my university.
9. I have very positive feelings about the engineering program's administration.
13. I like the attitude of this engineering program toward its students.

II. ACADEMIC EFFORT

2. In academics, I work as hard as I can work.
6. I focus on getting good grades in my major subject, which is essential to my career objective.
10. I work to succeed.
14. I do whatever it takes to succeed academically.

III. ACADEMIC CONFIDENCE

3R. I am having an academic problem in my engineering program.
7. I am absolutely certain that I will obtain an engineering degree here.
11R. I have to work harder in math than in any other courses.
15. It is not hard for me to get a B (3.0) average here.

IV. LACK OF ACADEMIC FIT/COMFORT

4R. I am not interested in my studies.
8R. I'm not sure this engineering program is for me.
12R. This engineering program needs a curriculum with more active hands-on, project-based course content.
16R. I wish there were more opportunities to learn research.

V. ENGINEERING FACULTY INTERACTIONS SCALE

I. FACULTY PERFORMANCE

1. The instructors in my engineering program are knowledgeable.
8R. The instructors do not ask questions that make me think.
15. My instructors often use interesting and innovative methods of teaching math and engineering.
22. The instructors are responsive when students ask questions.

II. FACULTY EXPECTATIONS OF STUDENTS

2. The instructors in this engineering program set high standards for students.
9. My instructors believe in the abilities of all students.
16. My instructors encourage me to excel.
23R. The instructors in this engineering program do not expect all students to succeed.

III. POSITIVE FACULTY-STUDENT INTERACTION

3. There is a faculty member that I particularly admire.
10R. I do not feel close to any faculty member.
17. There is a faculty member in my engineering program that I feel has been particularly helpful.
24. I have quite a bit of contact with engineering faculty members outside of class.

IV. SATISFACTION WITH FACULTY

4R. I am not satisfied with the teaching methods in my engineering courses.
11. I am pleased with the assistance from my professors in choosing courses.
18R. I am dissatisfied with the assistance from the engineering faculty in planning my future.
25. I am very satisfied with my courses because the instructors are inspiring.

V. FACULTY MENTORING OF STUDENT

5. I particularly enjoy being in the class of a particular professor because the professor thinks I am smart.
12. I am grateful that one of my instructors is teaching me how to succeed.
19. One of my professors has strongly encouraged me to go to graduate school in his/her field.

26R. I do not have a mentor (i.e. a teacher who is a good advisor) in my engineering program.

VI. FACULTY NEGLECT OF STUDENT(S)

6R. I have noticed that some instructor(s) tend to ignore me.

13R. I have an instructor who does not want to call on me, even though I know the answers.

20R. I feel that the engineering program instructors do not expect much from me.

27R. I do not feel that I receive enough praise from my instructors.

VII. EXPERIENCE OF RACISM

7. My instructors believe in the abilities of minority students.

14R. I feel that my engineering instructors sometimes use unfair grading procedures.

21. I don't detect much racism among the faculty here.

28R. Prejudice and racism are a problem here.

VI. STUDY HABITS SCALE

I. TIME SPENT STUDYING

2R. I do not like studying, so I tend to avoid it.

9. I try to increase study time.

13. When I find extra time, I study.

17. I go without sleep to do my studying.

23. I carefully schedule enough study time.

II. STRUCTURED STUDY HABITS

3. I study at same times most every day.

6. I study for certain number of hours and then stop.

10. I do not leave my studying to the last minute.

14. When I need help, I go for tutoring.

III. META-ORGANIZATION OF STUDYING

4. I envision the exam questions before taking an exam.

7R. I do not review class notes/information regularly.

11. I decide in advance what grade I want and then work toward it.

15. I study with the goal of improvement.

19. I develop many systems to help me with studying for coursework.

22. I spend a lot of time organizing coursework in addition to studying, i.e. making outlines, diagrams, charts, etc.

IV. MANAGING INTERFERENCE WITH STUDY

  5. I do not take phone calls when I am studying.
  8. Work interferes with my study time.
12R. My mind wanders when I study.
 16. I do not let people interrupt me when I study.
20R. When I try to study, other things always seem to come up.

V. STUDY WITH OTHERS

  1. I have a good study partner and we keep each other on track.
 18. I frequently study with friends or other students.
 21. I participate in a regular, organized study group with other students.
24R. I always study alone.

VII. ACADEMIC MANAGEMENT SCALE

I. TIME MANAGEMENT

  1. Proper time management will solve most problems in school.
  8. I know my academic priorities and plan my time accordingly.
 15. I control my time well.
 22. I can manage time well enough to get things done.
29R. Having enough time for everything is a big problem or me.

II. SCHEDULE MANAGEMENT

  2. I schedule my social life so that it does not interfere with study.
  9. I enjoy managing my daily schedule effectively.
 16. If I pay enough attention to my schedule, I can solve most of my problems.
 23. I plan my schedule far in advance.
30R. I rarely plan my day with a "To Do List."

III. MANAGERIAL PERSONALITY

 3R. I often feel incompetent.
10R. I lack discipline.
 17. I feel competent to manage my life.
24R. Inefficiency is a problem for me.
 31. I know how to get things done.

IV. LONG RANGE PLANNING

   4. I have written my own personal blueprint for success.

11R. I do not have specific goals to guide my activities.

18R. I am not making progress toward fulfilling my personal dream.

25R. I need to learn to plan ahead to avoid problems and crises.

  32. I can focus on a future goal for as long as necessary to attain it.

V. LIFE/STRESS MANAGEMENT

   5. I have learned how to handle stress.

  12. I know what is most important to my success in engineering.

19R. I have problems coping with anxiety and fear.

  26. I have learned how to maximize the conditions under which I function best.

  33. I get what I want in life.

VI. TEACHER MANAGEMENT

   6. To do well, I believe you should get to know your instructors.

  13. I assess a professor's attitudes before trying to get to know him or her.

  20. My strategy in engineering school is to find a special teacher, and establish a good relationship with that person.

  27. I think that the best way to get along with instructors is to learn from them.

  34. One of my strategies is to get to know a professor who will be able to write recommendations for me.

VII. PRO-ACTIVE (HELP-SEEKING) MANAGEMENT

   7. When I am having emotional problems, I get help.

  14. When I am having an academic problem, I get help from an instructor or tutor.

  21. There is a way to get help, even from teachers who are busy or unfriendly.

  28. I have encountered few problems in engineering school that I could not solve.

  35. I do not sit around and complain, I do something about a problem.

VIII. SOCIAL ADJUSTMENT SCALE

1. ADJUSTMENT TO PEOPLE, STUDENTS IN GENERAL

   1. In general, I like the people I have met in college.

7R.   Many of the other students here are not friendly.
14.   In general, I have strong positive feeling about the other students at my college.
20.   I have opportunities to collaborate with other students in my major.

II. ADJUSTMENT TO FRIENDS

2.   My friends here have helped my adjustment to college.
8.   I have a roommate.
15R.   I do not get along well with my roommate(s).
21.   I have made a wonderful best friend here.

III. SATISFACTION WITH DATING

3.   I am satisfied with the dating situation on campus.
9.   I have a boyfriend (or girlfriend), i.e. a dating partner.
16R.   I hardly ever date anyone from this school.
22R.   Problems with my "significant other" have made things difficult for me here.

IV. SENSE OF BELONGING

4R.   I feel excluded from many activities on campus.
10.   On this campus, I feel a sense of belonging.
17.   I have a good time here because I feel comfortable.
23R.   I am often very lonely here.

V. LEADERSHIP

5.   In small groups of people where I am comfortable, I am often looked to as the leader.
11.   What extra-curricular activities do you <u>actively</u> participate in?
12.   Indicate whether you hold a <u>leadership</u> position in any of them.
18.   I am a leader on campus.

VI. SOCIAL MANAGEMENT

6.   When I am having problems with depression, I talk to someone I feel close to.
13.   When I am feeling down, I never confide in friends.
19.   I enjoy being in groups of friends when we help each other with difficulties we are having.
24.   When I am having trouble, I talk to my family.
Note. Items 11 and 12 must be sequenced together.

IX. COMMUNITY INVOLVEMENT (IDENTITY)

I. COMMUNITY ORGANIZATION

1R. It isn't necessary that predominantly minority schools have a minority principal.
6R. Public services in the community neighborhoods need not be community controlled.
11. People should control their communities and what goes on in them.
21. Banks, stores and other businesses in a neighborhood ought to be owned and operated by people living there.

II. INVOLVEMENT/AGENCY

2. I am interested in what happens in the world.
7. I read newspapers.
12. I avidly follow news on television.
17. I am politically involved in community organizations.
22R. I do not usually attend lectures on current issues.

III. COMMUNITY IDENTITY INTEGRATION

3. Feeling able to compete against others is important to me.
4. Academic success is important to me.
9. Feeling that I have close friends is important to me.
14. I think about where I'll be five years from now.
19. Feeling socially successful is important to me.
24. Feeling part of the entire community is important to me.

IV. ETHNIC TRANSCENDENCE

5. I feel that people of different races are basically the same.
10. People of different races and ethnicities can and should be close friends.
15. People of different races and ethnicities have basically similar cultures
20. I believe there are prejudiced people of all races and ethnicities.
25. I have friends of many different groups.

V. ASSERTIVENESS FOR OTHERS

8. I feel able to stand up to others who behave with prejudice.
13. Fighting racism is important to me.

16. People should organize to improve their communities.
18. Working to build a strong community is important to me.
23. I intend to play a role in organizing communities.

X. SCIENTIFIC ORIENTATION

I. EARLY EXPOSURE TO SCIENCE

1. As a youngster, I liked to take things apart and put them back together.
6. My parents liked to give me puzzles, problems, and math games.
11. I always wanted to build thing and make things.
16. I participated in a FIRST Robotics program while I was in secondary school.

II. MATH COMPETENCE

2. Math is like a second language to me.
7. Ever since I can remember I liked numbers.
12. Being comfortable with math is one of my greatest assets in engineering school.
17R. I still struggle with mathematics.

III. PROBLEM SOLVER

3. I am a problem solver.
8. Give me a problem and I can find a way to solve it.
13. My parents taught me to solve problems.
18. From early on, I learned to organize things.

IV. KNOWABLE UNIVERSE

4. I believe that the universe is governed by laws.
9. Learning and knowledge are everything.
14. There is always a better way of doing things.
19R. Traditional ways are best for a good reason.

V. IMPROVE THE WORLD

5. I think a lot about how to make the world a better place.
10. Making good money is the primary concern to me.
15R. I never think about improving the human condition.
20. Engineering is of great value to society.

XI. LEARNING STYLE INVENTORY

Learning Styles (Felder & Silverman, 1988; Felder et al., 2002) have been the subject of much research aimed at understanding the personalities that best thrive in engineering as well as how to change the curriculum to accommodate a wider array of learning styles. Adapted from Kolb's (1984) Learning Styles, the 8-pronged measure was used to assess Active, Reflective, Visual, Auditory, Intuitive, Sensing, Sequential, and Global learning styles. Note that inductive and deductive dimensions were eliminated from the inventory by Felder and Silverman (1988). The 44 items for this scale can be found at www.webtools.ncsu.edu/learningstyles/

1. I understand something better after I

   _____ (a)   try it out
   _____ (b)   think it through

2. I would rather be considered

   _____ (a)   realistic
   _____ (b)   innovative

3. When I think about what I did yesterday, I am most likely to get

   _____ (a)   a picture
   _____ (b)   words

4. I tend to

   _____ (a)   understand details of a subject but may be fuzzy about its overall structure
   _____ (b)   understand the overall structure but may be fuzzy about details

5. When I am learning something new, it helps me to

   _____ (a)   talk about it
   _____ (b)   think about it

6. If I were a teacher, I would rather teach a course

   _____ (a)   that deals with facts and real life situations
   _____ (b)   that deals with ideas and theories

7. I prefer to get new information in

   _____ (a)   pictures, diagrams, graphs, or maps
   _____ (b)   written directions or verbal information

8. Once I understand

   _____ (a)   all the parts, I understand the whole thing
   _____ (b)   the whole thing, I see how the parts fit

9. In a study group working on difficult material, I am more likely to

 _____ (a) jump in and contribute ideas
 _____ (b) sit back and listen

10. I find it easier

 _____ (a) to learn facts
 _____ (b) to lean concepts

11. In a book with lots of pictures and charts, I am likely to

 _____ (a) look over the pictures and charts carefully
 _____ (b) focus on the written text

12. When I solve math problems

 _____ (a) I usually work my way to the solutions one step at a time
 _____ (b) I often just see the solutions but then have to struggle to figure out the steps to get to them

13. In classes I have taken

 _____ (a) I have usually gotten to know many of the students
 _____ (b) I have rarely gotten to know many of the students

14. In reading nonfiction, I prefer

 _____ (a) something that teaches me facts or tells me how to do something
 _____ (b) something that gives me new ideas to think about

15. I like teachers

 _____ (a) who put a lot of diagrams on the board
 _____ (b) who spend a lot of time explaining

16. When I am analyzing a story or novel

 _____ (a) I think of the incidents and try to put them together to figure out the themes
 _____ (b) I just know what the themes are when I finish reading and then I have to go back and find the incidents that demonstrate them

17. When I start a homework problem, I am more likely to

 _____ (a) start working on the solution immediately
 _____ (b) try to fully understand the problem first

18. I prefer the idea of

 _____ (a) certainty
 _____ (b) theory

19. I remember best

    _____ (a) what I see
    _____ (b) what I hear

20. It is more important to me that an instructor

    _____ (a) lay out the material in clear sequential steps
    _____ (b) give me an overall picture and relate the material to other subjects

21. I prefer to study

    _____ (a) in a study group
    _____ (b) alone

22. I am more likely to be considered

    _____ (a) careful of the details of my work
    _____ (b) creative about how to do my work

23. When I get directions to a new place, I prefer

    _____ (a) a map
    _____ (b) written directions

24. I learn

    _____ (a) at a fairly regular pace. If I study hard, I'll "get it"
    _____ (b) in fits and starts. I'll be totally confused and then suddenly it "clicks"

25. I would rather first

    _____ (a) try things out
    _____ (b) think about how I'm going to do it

26. When I am ready for enjoyment, I like writers to

    _____ (a) clearly say what they mean
    _____ (b) say things in creative, interesting ways

27. When I see a diagram or sketch in class, I am most likely to remember

    _____ (a) the picture
    _____ (b) what the instructor said about it

28. When considering a body of information, I am more likely to

    _____ (a) focus on details and miss the big picture
    _____ (b) try to understand the big picture before getting into the details

29. I more easily remember

_____ (a) something I have done
_____ (b) something I have thought a lot about

30. When I have to perform a task, I prefer to

_____ (a) master one way of doing it
_____ (b) come up with new ways of doing it

31. When someone is showing me data, I prefer

_____ (a) charts or graphs
_____ (b) text summarizing the results

32. When writing a paper, I am more likely to

_____ (a) work on (think about or write) the beginning of the paper and progress forward
_____ (b) work on (think about or write) different parts of the paper and then order them

33. When I have to work on a group project, I first want to

_____ (a) have "group brainstorming" where everyone contributes ideas
_____ (b) brainstorm individually and then come together as a group to compare ideas

34. I consider it higher praise to call someone

_____ (a) sensible
_____ (b) imaginative

35. When I meet people at a party, I am more likely to remember

_____ (a) what they looked like
_____ (b) what they said about themselves

36. When I am learning a new subject, I prefer to

_____ (a) stay focused on that subject, learning as much about it as I can
_____ (b) try to make connections between that subject and related subjects

37. I am more likely to be considered

_____ (a) outgoing
_____ (b) reserved

38. I prefer courses that emphasize

_____ (a) concrete material (facts, data)
_____ (b) abstract material (concepts, theories)

39. For entertainment, I would rather

_____ (a) watch television
_____ (b) read a book

40. Some teachers start their lectures with an outline of what they will cover. Such outlines are

_____ (a) somewhat helpful to me
_____ (b) very helpful to me

41. The idea of doing homework in groups, with one grade for the entire group

_____ (a) appeals to me
_____ (b) does not appeal to me

42. When I am doing long calculations

_____ (a) I tend to repeat all my steps and check my work carefully
_____ (b) I find checking my work tiresome and have to force myself to do it

43. I tend to picture places I have been

_____ (a) easily and fairly accurately
_____ (b) with difficulty and without much detail

44. When solving a problem in a group, I would be more likely to

_____ (a) think of the steps in the solution process
_____ (b) think of possible consequences or applications of the solution in a wide range of ideas

7 The College Adjustment of Black and Hispanic Students in Engineering

Summary

The general literature on minority students in college finds similar issues of adjustment for African American and Hispanic students. This proved true for these students in engineering, as well. An online survey of the undergraduate engineering experience was completed by 260 Black students (African American, African and Caribbean) and 368 Hispanic students to assess success factors for minorities in engineering. The survey included measures of Minority Engineering Program participation, college adjustment, and background. The results showed that for both groups of students, a combination of academic skill sets, satisfaction with the engineering environment, and participation in critical program components, such as internships, contributed most to three measures of "success" in a series of statistical analyses.

Black students participated more in MEP components, and their success depended more on program involvement. Matriculation in one of the Historically Black Colleges and Universities (HBCUs) contributed significantly to enhanced 6-year graduation rates for African American students. There was evidence that making a happy adjustment to the college environment was more predictive for these students, which may help explain why HBCUs with strong engineering programs constitute such a significant success factor for them. There were few gender differences among African Americans.

Compared to Black students, Hispanic students reported a stronger scientific orientation, while their success relied more heavily on managerial skill and interpersonal interactions of several kinds—with peers, in study groups, and with faculty. Hispanic students appeared to participate in MEP programs when their grades were low. Attending an Hispanic-Serving Institution (HSI) had no effect on engineering success for these students. Among Hispanics, female students participated more in MEP program offerings, and were more outgoing.

There were only minor differences in test scores and academic performance between groups and genders. The findings argue for strengthened academic skills including test scores, and increased access to and participation in research and particularly internships for both groups of underrepresented students. Because MEP programs appear to enable a more positive adjustment to the engineering environment, minority students would do well to maximize their participation in them.

Introduction

This study of success factors for minorities in engineering has entailed three different phases. It has explored institutional data, focus group data, and an online survey. Thus far, the analyses have established little difference in minority and non-minority retention to graduation rates among students of equal ability, and that exposure to hands-on immersion in the work of engineering, as in problem or project-based instruction, research, and internships, was associated with multiple measures of minority student "success." Furthermore, minorities in engineering reported better adjustment to the engineering school environment than non-minority students when there were differences. Greater minority participation in program offerings may well be the reason for their better adjustment, and account for there being very little difference in minority and non-minority academic performance, despite lower minority test scores. The information gleaned from the analyses so far would be incomplete without some understanding of the differences in success factors between African American and Latino/a students, as well as differences by gender. This ethnic comparison also affords an opportunity to examine the influence of Minority-Serving Institutions on minority engineering students.

Background

Both African American and Hispanic students are underrepresented in engineering, and are thus groups targeted for corrective efforts, as well as an increased need to understand the factors that encourage minority talent. Several studies have reported high raw dropout rates in engineering for African Americans up to 67% (Landis, 2005; Morning & Fleming, 1994; Yoder, 2012). Chubin, Donaldson, Olds, and Fleming (2008), as well as Brown, Morning, and Watkins (2005), determined that African American engineering students were at much greater risk of dropping out than other students. Recent data indicates that while African Americans are 12.3% of the population, they earn only 4.0% of the undergraduate degrees, 3.6% of engineering workforce (NACME, 2014a). Furthermore, they represent only 3% of the engineering faculty. Compared to the undergraduate retention to graduation rate of 72.8% for Asian Americans and 63.7% for White Americans, only 31.2% of African Americans are retained to graduation. Yoder (2012) reports similar figures, but with 38.6% of African Americans being retained to graduation.

Hispanic students have shown the most progress in engineering participation, but remain underrepresented (NACME, 2014b). They constitute 17% of the population, but 8.6% of the undergraduate degrees, 6.3% of the engineering workforce, and only 3.7% of engineering faculty. Most the advanced degrees earned come from two institutions: Puerto Rican universities and the University of Texas, El Paso. While Latinas were almost nonexistent in the pipeline in 1997, they now constitute 40% of the graduates from University of Puerto Rico. However, the Latina percentage of undergraduate and master's degrees has remained stagnant for the last 10 years. Yoder (2012) reports that while the 6-year graduation rate of Asian American engineering students was 66.5%, and Caucasian students was 59.7%, the rate for Hispanics in engineering was 44.4%—about 6 percentage points higher than for African American students in the same study.

The literature on minorities in engineering is sparse, perhaps owing to their small numbers (Hackett, Betz, Casas, & Rocha-Singh, 1992; Lewis, 2003). However, there is a small but growing number of studies investigating success and adjustment factors for African American and Latino/a students, although many studies still lump the two groups together or combine minority students from all of the STEM disciplines. Most studies do find that the two groups have similar adjustment issues. The literature on minority students in college in general discusses minority student adjustment at Predominantly White Institutions and suggests that both African American and Latino/a students experience a major period of transition, which requires the use of coping strategies for establishing supportive relationships and maintaining academic performance, emotional and physical well-being (Chiang, Hunter, & Yeh, 2004; Cole, 2008; Solberg & Viliarreal, 1997; Tomlinson-Clarke, 1998). Constantine, Chen, and Ceesay (1997) report that African American and Latino/a students share similar college adjustment issues in the areas of academics, finances, personal/family relationships, and stress management (see also Constantine, Wilton, & Caldwell, 2003). However, it has been suggested that Hispanic students have fewer ethnically related adjustment issues because they are less visible than African Americans, and that Black students are more likely than Latino/as to feel alienated and perform poorly (Ancis, Sedlacek, & Mohr, 2000; Oliver, Rodriguez, & Mickelson, 1985; Suarez-Balcazar, Orellana-Damacela, Portillo, Rowan, & Andrews-Guillen, 2003). Further, there is allegedly greater resistance to the acceptance and assimilation of Blacks (Hacker, 2010). The general college adjustment literature on HBCUs and HSIs indicates that while African American students fare far better in HBCUs than in White institutions (Allen, 1992; Constantine, 1995; Fleming, 1984; Kim, 2002; Kim & Conrad, 2006), that Hispanic students show fewer differences in HSIs compared to White institutions (Gasman, 2008; Bridges, Kinzie, Laird, & Kuh, 2008). Part of the reason for this institutional impact difference may be that while HBCUs were chartered to improve the educational position of African Americans, HSIs were so designated by virtue of the percentages of Hispanic students over

25% rather than possessing a mission to serve these students (Conrad & Gasman, 2015; Contreras, Malcolm, & Bensimon, 2008).

In the literature on minorities in science, similar issues for both groups have also been reported, such as problems with retention in engineering and the nurturing of intelligence and critical thinking abilities (Fleming, Garcia, & Morning, 1996; Fleming & Morning, 1998; Morning & Fleming, 1994), but the results were not disaggregated by ethnicity. Wolfe, Powell, Schlisserman, and Kirshon (2016) found that minority students in engineering, as well as women, reported more teamwork problems than other students such that they were excluded from the most meaningful work with consequences for optimal learning. Person and Fleming (2012) found that Black students in STEM fields performed worse than other minorities in Predominantly White Institutions (PWIs), but this ethnic factor made no difference in Black colleges or community colleges.

A number of studies find that engineering students' positive interactions with faculty contribute to successful outcomes (Amelink & Meszaros, 2011; Hackett et al., 1992; Vogt, 2008), but are especially critical for Black and Latino/a students, perhaps due to alleged social capital disadvantages (Martin, Simmons, & Yu, 2013; Cole & Espinoza, 2008; Hurtado et al., 2011). While routine interactions with faculty concerning grades may not lead to an academic advantage (Cole & Espinoza, 2008), faculty mentorship and research with faculty members generally benefit minority students (Barlow & Villarejo, 2004; Kim & Sax, 2009; Maton & Hrabowski, 2004). Black and Latino/a students may resist establishing contact with faculty who are not same-ethnicity, but interactions with peers and staff may provide a substitute for this form of support (Dika, Pando, & Tempest, 2016).

Previous institutional statistics from this study found that Hispanic students perform somewhat better than African American students, given similar ability levels (see Chapter 3). Also, results from focus group minisurveys (see Chapter 5) found few ethnic differences between the two groups, but other studies suggest that psychological resilience is a predictive factor in success for African Americans, suggesting that differences in history and conditioning may play a role in their adjustment to the engineering environment (e.g., McPherson, 2017). Moore, Madison-Colmore, and Smith (2003) have reported special challenges facing Black males in engineering, requiring a greater determination to persist and succeed that they label the "prove-them-wrong" attitude (see Moore, 2006).

Without race-based pressures, African American students appear to thrive in HBCUs, according to reports that identify HBCUs among the top producers of Blacks in engineering (Malcolm-Piquex & Malcolm, 2015; Ransom, 2015; Reichert & Absher, 1997). Lent et al. (2005) determined that compared to non-Black students in PWI engineering schools, Black students in an HBCU engineering schools reported stronger self-efficacy, higher outcome expectations, greater technical interest and social support. Flowers (2014) reported that finding mentors was a problem for Black males in White engineering schools, but not in Black engineering schools. From case studies, Bonner (2014) theorizes that White schools may support the academic goals of Black males in STEM disciplines, but Black schools are also able to support the personal and cultural aspects of the individual outside of the classroom.

In sum, African American and Hispanic students share similar adjustment issues, but there is some evidence that Hispanic students may fare somewhat better. However, HBCUs confer an educational advantage for Black students, although the same cannot be said for HSIs and Hispanic students.

Purpose of the Study

This analysis compares the patterns of program participation and adjustment of African American and Hispanic students from an online survey of undergraduate engineering students.

It raises the following questions:

1. Are there observed ethnic differences in patterns of program participation, college adjustment, or background?
2. Are there observed ethnic differences in factors contributing to enhanced GPA?
3. Are there different ethnic factors related to effective schools; that is, to schools more effective in graduating minority students?
4. Are there observed ethnic differences in factors contributing to retention or longevity in engineering schools measured by freshman-senior differences?
5. What are the implications of observed differences in the adjustment and success patterns for student success and program effectiveness?

Method

An online survey including 11 measures of program participation and college adjustment were administered via Survey Monkey to students attending 18 institutions. The analysis compared ethnic differences in the study measures, and within each group determined the most important variables predicting three measures of "success:" overall GPA; matriculation in engineering schools more effective in graduating minorities; and in freshman-senior differences suggesting better retention. For this analysis that focuses solely on Black and Hispanic students, the role of Minority-Serving Institutions could also be investigated. Of the institutions represented, three were HBCUs and five were designated as HSIs with at least 25% Hispanic enrollment.

Table 7.1 describes the study sample of Black and Hispanic students. Within a total population of 632 minority students (excluding Native

Table 7.1 Study Population

Ethnicity	All n = 628	Male 60.2% n = 378	Female 39.8% n = 250	GPA 3.03 SD = 0.49	Test Score Percentile 72.1 SD = 21.7
Black	41.4% *(260)*	52.7% (137)	47.3% (123)	3.01 SD = 0.51	74.7% SD = 20.7
Hispanic	58.5% (368)	65.5% (241)	34.5% (127)	3.05 SD = 0.49	70.3% SD = 22.3

Note. Number of students in parentheses. Black students include 225 African American, and 35 African/ Caribbean students.

Americans), 260 (or 41.4%) were Black (including 225 African American and 35 African or Caribbean students); 368 (or 62.1%) were Hispanic. Male students numbered 362 (or 61.4%) and female students numbered 229 (or 38.6%) of the group. There was a higher percentage of male students among Hispanics compared to African Americans (65.5% vs. 54.6%), and likewise a higher percentage of females among African Americans than among Hispanics (45.3% vs. 34.5%). While the study attempted to recruit Native American students, there were only four in the survey. They were not included in this analysis. Also not included were biracial students.

The measures, which included program involvement and college adjustment, were adapted from the evaluation of nine different programs in various STEM fields as well as studies of student development in college (Felder & Silverman, 1988; Felder, Felder, & Dietz, 2002; Fleming, 1984, 2001, 2012; Hardy, 1974; Moos, 1979; Nasim, Roberts, Harrell, & Young, 2005; Nixon & Frost, 1990; Pearson, 2005; Person & Fleming, 2012; Ramseur, 1975; Sedlacek, 1998, 2004). Measures of college adjustment were grouped into two categories: those specific to college adjustment dimensions—College Adjustment, Study Habits, Academic Adjustment, Social Adjustment, Faculty Interactions, Comfort; and those measuring personal orientations that might influence or interact with college adjustment—Learning Styles, Community Identity, and Scientific Orientation. Throughout the report, they are collectively referred to as college adjustment measures. The details of instrumentation and analysis are presented in Appendices 1–3 of Chapter 6.

Comparing African American and Hispanic Students

The average test score percentile was 72.1%; test score percentiles were significantly higher for Black students (74.7% vs. 70.3%; see Table 7.2). Average GPA was 3.03, with non-significant differences between the groups—2.99 for African Americans and 3.05 for Hispanics. With test scores controlled, Black GPA was 2.98 compared to 3.05 for Hispanic students. While Latino/a test scores were slightly lower, their grades were

Table 7.2 Black and Hispanic Engineering Student Differences in Program Participation, College Adjustment, and Background Measures, Controlling for Test Score Percentile

STUDY VARIABLE	Black Students n = 260	Hispanic Students n = 368	F/p	Effect Size (Eta)
DEMOGRAPHIC & BACKGROUND VARIABLES:				
GPA	3.01 (0.03)	3.05 (0.03)	ns	na
GPA, With Test Score Percentile	2.99 (0.03)	3.05 (0.03)	ns	na
Test Score Percentile	74.7 (1.34)	70.3 (1.13)	6.32*	.010
SES: Parents' Years of Education	2.71 (0.06)	2.31 (0.05)	31.7***	.048
Average Hours Worked/Week	19.1 (1.03)	17.2 (0.81)	ns	na
Have Job Outside Program	42.3%	49.2%	ns	na
Level of Scholarship Support	2.14 (0.04)	2.22 (0.03)	ns	na
NACME Scholar	53 25.6% n = 207	48 17.2% n = 279	Fisher = *	na
RAW PROGRAM PARTICIPATION PERCENTAGES:				
Middle School STEM Programs	31.5%	23.1%	Fisher = *	
High School STEM Programs	43.5%	35.1%	Fisher = *	
Summer Bridge	22.3%	27.2%	ns	
Freshman Orientation Courses	54.6%	47.3%	ns	
Tutoring Programs	64.6%	50.5%	Fisher = ***	
Supplemental Instruction	60.4%	42.4%	Fisher = ***	
Study Groups	39.6%	47.0%	ns	
Peer Mentor Programs	24.2%	24.2%	ns	
Faculty Mentor Programs	26.2%	31.3%	ns	
Research With a Faculty Member	18.1%	17.4%	ns	
Corporate Internships	35.4%	17.9%	Fisher = ***	

(*Continued*)

Table 7.2 (Continued)

STUDY VARIABLE	Black Students	Hispanic Students	F/p	Effect Size (Eta)
PROGRAM PARTICIPATION SCALES:				
Program Participation Scale	30.4 (0.61)	27.7 (0.51)	11.27***	.018
Middle School STEM Program Participation Scale	2.49 (0.13)	2.07 (0.11)	6.32*	.010
High School STEM Program Participation Scale	3.04 (0.14)	2.67 (0.12)	3.81~*	.006
Summer Bridge Participation Scale	2.18 (0.14)	2.23 (0.12)	ns	na
Freshman Orientation Course Participation Scale	3.26 (0.15)	3.06 (0.09)	ns	na
Tutoring Participation Scale	3.99 (0.14)	3.26 (0.12)	15.66***	.024
Supplemental Instruction Participation Scale	3.78 (0.15)	2.96 (0.12)	18.38***	.029
Study Group Participation Scale	3.09 (0.16)	3.24 (0.13)	ns	na
Peer Mentoring Participation Scale	2.13 (0.13)	2.09 (0.11)	ns	na
Faculty Mentoring Participation Scale	2.12 (0.14)	2.43 (0.12)	ns	na
Research Participation Scale	1.76 (0.12)	1.85 (0.09)	ns	na
Internship Participation Scale	2.67 (0.13)	1.88 (0.11)	22.09***	.034
COLLEGE ADJUSTMENT SCALES:				
College Adjustment Scale	73.7 (0.94)	74.9 (0.76)	ns	na
Comfort Scale	43.9 (0.55)	44.6 (0.45)	ns	na
Academic Adjustment Scale	59.1 (0.53)	60.1 (0.43)	ns	na
Engineering Faculty Interactions Scale	119.0 (1.38)	121.5 (1.13)	ns	na
Academic Management Scale	119.5 (1.41)	120.6 (1.15)	ns	na
Study Habits Scale	66.14 (0.58)	63.51 (0.64)	ns	na
Social Adjustment Scale	104.9 (1.21)	105.2 (0.98)	ns	na
Scientific Orientation Scale	66.5 (0.71)	69.6 (0.59)	11.29***	.018
Community Identity Scale	99.4 (0.57)	99.4 (0.76)	ns	na

(*Continued*)

Table 7.2 (Continued)

STUDY VARIABLE	African American Students	Hispanic Students	F/p	Effect Size (Eta)
LEARNING STYLES:				
Active	6.09 (0.16)	6.27 (0.13)	ns	na
Reflective	4.90 (0.16)	4.73 (0.13)	ns	na
Sensing	6.26 (0.14)	6.62 (0.12)	3.96*	.006
Intuitive	4.74 (0.14)	4.38 (0.12)	3.96*	.006
Visual	7.77 (0.14)	7.87 (0.12)	ns	na
Auditory	3.22 (0.14)	3.13 (0.12)	ns	na
Sequential	6.35 (0.13)	6.55 (0.11)	ns	na
Global	4.74 (0.14)	4.46 (0.10)	ns	na

Note. Standard error of the mean in parenthesis. Effect size (partial *Eta* squared) given for significant results.
* $p < .05$
** $p < .01$
*** $p < .001$

slightly higher than for African American students. African American students reported higher parental education levels than Hispanic students and were more likely to be NACME Scholars. There were no differences in average levels of scholarship support or time spent working at a job.

On program participation measures, there were five differences, and Black students scored higher on all of them. They reported greater overall Program Participation, greater Middle School STEM Program Participation, and more participation in Tutoring, Supplemental Instruction, and Internships. Internship participation has emerged as a critical experience for engineering students and an important success factor; the greatest ethnic divergence was in internship participation. African American students reported a 35.4% participation in industry internships, compared to 17.9% of Hispanic students (and compared to 43% of non-minority students).

On college adjustment measures, there was only one significant difference—Hispanic students reported a stronger Scientific Orientation than African American students. An investigation of the five subscales of this measure indicated that Hispanic students scored significantly higher on three of them: the strongest difference was in the self-assessment of Math Competence; followed by self-assessment as a Problem-Solver; and belief in a lawful Knowable Universe. On learning styles, Hispanic students were more Sensing while Black students were more Intuitive.

In short, although African Americans participate more in MEP programs, Hispanic students report a stronger scientific orientation.

Adjustment of Black Engineering Students

Enhancing Academic Performance

When the study measures were entered into a regression equation (Table 7.3), the most predictive factors for academic success were: Internship Involvement; Research Participation; and Academic Adjustment. These factors, however, accounted for only 8.8% of the variance in grades.

It appears that access to internships and research, combined with making a good adjustment to the academic milieu, provides the recipe for Black student academic success.

Effective Schools

Table 7.3 shows that in a regression equation along with test scores and background variables, three measures were significantly associated with more effective schools: attending an HBCU; Internship Involvement; and better College Adjustment scores, accounting for 34.6% of the variance in effective school rankings.

In previous analyses of minority students, it was not possible to account for adequate amounts of variance in the effective school dimension. For African American students, inclusion of an HBCU variable effectively resolves that problem; its inclusion enables the study to account for about one-third of the variance in effective schools. The additional variable of Internship Involvement continues to support the idea that exposure to the work of engineering is an important component in student success, while the College Adjustment variable indicates that Black students effect a better adjustment in more effective schools.

Freshman-Senior Differences

Among African American students, GPA did not change with academic status. Table 7.3 shows that the most important differences were that seniors scored higher on Internship Involvement, Research Participation, Global Learning Styles, and Faculty Mentoring. Seniors were lower on test scores, Peer Mentoring, and Auditory Learning Styles. These factors accounted for 20.3% of the variance in academic status differences.

Success Factors

Table 7.4 presents a factor analysis of the measures that were positively associated with the three dependent measures of "success." It shows

Table 7.3 Regression Analyses of Measures of Program Participation and College Adjustment on Success Variables for Black Engineering Students (n = 260)

Dependent Variables	Independent Variables	t	R/R²	F
GPA:				
	Internship Involvement Scale	3.02**		
	Research Participation Scale	2.15*		
	Academic Adjustment Scale	2.32*	R = .296 R^2 = .088	8.19***
Effective Schools Rankings:				
	Attending an HBCU	9.37***		
	Internship Involvement Scale	3.74***		
	College Adjustment Scale	2.05*	R = .588 R^2 = .346	36.91***
Academic Status; i.e., Freshman-Senior Differences:				
	Internship Involvement Scale	4.51***		
	Research Participation Scale	2.98**		
	Global Learning Styles	2.03*		
	Faculty Mentor Participation Scale	1.98*		
	Test Score Percentile	−3.19**		
	Peer Mentor Participation Scale	−2.84**		
	Auditory Learning Styles	−2.52*	R = .451 R^2 = .203	9.26***

Note. Positive predictors in italics.
* $p < .05$
** $p < .01$
*** $p < .001$

Table 7.4 Factor Analysis of Success Factors for Black Engineering Students

Factors	Factor Loadings
I: Academic and College Adjustment	
Academic Adjustment Scale	.863
College Adjustment Scale	.852
II: Internship Participation vs. HBCU	
Global Learning Styles	.797
Attending an HBCU	−.632
Internship Involvement Scale	.459
III: Research and Faculty Involvement	
Research Participation Scale	.775
Faculty Mentor Participation Scale	.767

three factors. The first, Academic and College Adjustment, is composed of the Academic Adjustment Scale and the College Adjustment Scale. The second is Program Participation vs. HBCU Attendance. On the one hand, it suggests that Black student participation in Internships, along with Global (i.e., theoretical) Learning Styles is a pathway to success in engineering. On the other hand, it suggests that attendance at an HBCU constitutes a pathway in its own right. The third factor of Research Participation and Faculty Involvement implies that the mentoring gained through research participation constitutes a powerful success mixture. This analysis succinctly describes what works for African American students—a satisfactory adjustment, immersion in critical program components of internships and/or research, or attending an HBCU. Using all scales and subscales in a factor analysis did not improve the interpretation and did not improve prediction in the foregoing analyses.

The Influence of Gender Among Black Students

There were no gender differences in program participation or their components, nor in college adjustment scales. On personal orientation scales, African American females were more Global in learning styles, while males were more Sequential. There were no interactions by gender as a function of matriculating in more effective institutions, and none as a function of freshman-senior differences. When all scales and subscales were utilized, 37.6% of the variance in male GPA, and 41.4% of the variance in female GPA could be accounted for.

Adjustment of Hispanic Engineering Students

Enhancing Academic Performance

Better grades for Hispanic students, as has been seen for others, depends on a combination of higher test scores, sufficient academic managerial skills, and a happy adjustment to the academic milieu. Table 7.5 shows that when the study measures were entered into a regression equation, the positive factors were test scores, Academic Management, and Academic Adjustment. Factors associated with lower grades were Summer Bridge Participation, and surprisingly, Scientific Orientation. These variables accounted for 15.9% of the variance in grades. Although a familiar set of factors was associated with higher GPA for these as well as other students, the negative effect on grades of a higher reported scientific orientation is puzzling. Hispanic students scored higher than Black students on this measure. Further, the Early Exposure to Science subscale bore the strongest negative relationship to GPA. No ready explanation presents itself.

Effective Schools

In previous reports from this study, effective schools were those in which students made the most use of MEP program components. The same was true for Hispanic students in schools more effective in graduating them in six years. However, while Black students in HBCUs were conferred a graduation advantage, the same cannot be said for Hispanic students in HSIs.

Table 7.5 shows that the factors most strongly associated with effective schools were Freshman Orientation Course Participation, Engineering Faculty Interactions, Internship Involvement, and Summer Bridge Participation. These factors accounted for 22.1% of the variance in effective school rankings.

Positive faculty interactions, and program participation, then, appear to be the hallmarks of effective engineering schools for Hispanic students—as for others. Yet there were several other noteworthy trends. Although the HSI designation had no positive effect on graduation rates for these students, it is evident that positive student-faculty interactions play a significant role in higher graduation rates. This result leads to the unanswered question: could happier relationships with engineering faculty substitute for an ethnically comfortable engineering school environment?

Freshman-Senior Differences

As for other students, Hispanic students who survive to the upperclass years in engineering were characterized by greater program participation of various kinds and better adjustment of certain kinds.

The regression equation showed that senior factors were Social Adjustment, Study Group Participation, Research Participation, and Internship Involvement (Table 7.5). Freshman factors were Engineering Faculty Interactions, Peer Mentoring, Freshman Orientation Course Participation, and Comfort. These variables accounted for 25.8% of the variance; they also supersede test scores.

Table 7.5 Regression Analyses of Measures of Program Participation and College Adjustment on Success Variables for Hispanic Engineering Students (n = 368)

Dependent Variables	Independent Variables	t	R/R^2	F
GPA:				
	Test Score Percentile	.5.22***		
	Academic Management Scale	3.09**		
	Academic Adjustment Scale	2.49*		
	Summer Bridge Participation Scale	−2.38*		
	Scientific Orientation Scale	−2.17*	$R = .399$ $R^2 = .159$	13.69***
Effective School Rankings:				
	Freshman Orientation Course Participation Scale	4.18***		
	Engineering Faculty Interactions Scale	3.76***		
	Internship Involvement Scale	3.67***		
	Summer Bridge Participation Scale	2.97**	$R = .417$ $R^2 =. 174$	118.26***
Academic Status; i.e., Freshman-Senior Differences:				
	Social Adjustment Scale	4.07***		
	Study Group Participation Scale	3.90***		
	Research Participation Scale	3.76***		
	Internship Involvement Scale	2.58*		
	Engineering Faculty Interactions Scale	−3.83***		
	Peer Mentor Participation Scale	−3.44**		
	Freshman Orientation Course Participation Scale	−3.08**		
	Comfort Scale	−2.78**	$R = .508$ $R^2 = .258$	15.59***

Note. Positive predictors in italics.
* $p < .05$
** $p < .01$
*** $p < .001$

In the academic literature, there is little as important as faculty interactions for student well-being. Indeed, for Hispanic students, faculty interactions were a critical factor in higher GPA. Yet, in this analysis, senior engineering students reported less favored relationships with their faculty than underclassmen. When all subscales of Faculty Interactions were considered, two subscales were lower among seniors: Faculty Expectations and Faculty Racism. Seniors were less likely to think that their faculty held high expectations for them, but they apparently did not attribute this to racism. Other senior factors imply positive interactions, such as social adjustment, participation with others in study groups, research, and internships. The question still is: can social and academic group interactions substitute for a lessened rapport with faculty? To at least a certain extent, it seems so.

Success Factors

Table 7.6 presents the factor analysis of program and summary measures positively associated with the three dependent variables in this study. The four factors were:

1. *Academic and Interactive Adjustment.* This includes academic adjustment and management skills, as well as faculty and social interactions. It describes a full range of adjustment measures.
2. *Entry Level Program Participation.* This is composed of entry level programs such as Summer Bridge and Freshman Orientation Courses. Participation in these programs was associated with lower test scores.

Table 7.6 Factor Analysis of Success Factors for Hispanic Engineering Students

I: Academic and Interactive Adjustment	
Academic Adjustment Scale	.849
Engineering Faculty Interactions Scale	.829
Academic Management Scale	.791
Social Adjustment Scale	.680
II: Entry Level Program Participation vs. Test Score Percentile	
Sumer Bridge Participation Scale	.806
Test Score Percentile	−.621
Freshman Orientation Course Participation Scale	.614
III: Interactive Program Involvement	
Study Group Participation Scale	.759
Research Participation Scale	.537
IV: Internship Involvement	
Internship Involvement Scale	.842

3. *Interactive Program Involvement.* This is composed of participation in study groups and faculty-guided research.
4. *Internship Involvement.* This program component occupies a factor of its own.

Since there has been a concern with the amount of variance accounted for in the various regression analyses, all scales and subscales were utilized for this group of students. The variance estimate for GPA increased from 15.9% to 24.1%; for effective schools, from 22.1% to 28.6%; and for freshman-senior differences, from 25.8% to 38.2%. The resulting factors and interpretation are similar to those using summary scales.

The Influence of Gender Among Hispanic Students

While there were few gender differences among African American students, there were more among Hispanic engineering students. Hispanic females were more involved in MEP programming, with faculty, and with other students, and exhibited a broader community identity.

Hispanic women reported higher Program Participation, including more High School Program participation. They reported higher scores on College Adjustment, Faculty Interactions, Academic Management, Social Adjustment, and Community Identity. It could be said that Hispanic females were characterized by better adjustment to the engineering environment, with adjustment advantages that include more MEP participation, better college satisfaction, better academic skills, and more rapport with people. Their greater rapport spans interactions with other students, faculty, and community.

Despite the gender differences observed for the study measures, there were few differences observed in patterns of prediction to the "success measures" used. For each gender, an Academic and Social Management Factor emerged. For males, this factor was weighted more strongly toward academic skills, while for females it was weighted more heavily toward social interactions with peers and faculty. This appears to be consistent with the generally more outgoing character of Hispanic females. For males only, test scores constituted a separate success factor. It may be that for females, their more interactive nature, combined with academic managerial skill, supersede the effect of test scores.

Conclusion

The general literature on minority students in college finds similar issues of adjustment for Black and Hispanic students. Likewise, this study of success factors has found few ethnic differences between the two groups of students. While Black students had higher test scores, both groups of students exhibited similar overall GPAs with or without test score controls. Any differences in performance were minor and non-significant. There were, however, several differences worthy of note.

First, Black students participate more in MEP program offerings. These programs include tutoring, supplemental instruction, and particularly industry internships. While 35.4% of Black students participate in internships, only 17.9% of Hispanic students report such participation. Recall that non-minority students participate in internships to a much greater degree than minority students; i.e., 43%. Furthermore, internship involvement routinely emerges as a critical success factor. Students who participate in them achieve higher grades, are more often in effective schools, and survive to the upperclass years.

Second, Hispanic students reported a stronger scientific orientation, compared to Black students. The subscales of this measure supporting higher scores for Hispanic students included math competence, being a problem solver, and belief in a lawful, knowable universe. On the other hand, despite scoring higher on this scale, scientific orientation proved to be a negative factor in grades for Hispanic students. Specifically, the early exposure to science subscale proved to be the most important negative factor on this scale. While no ready explanation for these results presents itself, it does pique curiosity about the significance of scientific orientation for both groups of students. This new, experimental scale may deserve greater attention.

Third, at first blush, the success factors for both groups of students appear to be quite similar. They are comprised of a happy adjustment to the engineering environment, academic skills, and participation in various MEP program offerings. Ability, measured by test scores, figures prominently in minority student success. Yet note these differences. Black students' success depends more on MEP program involvement. Hispanic students' success relies more on managerial skill and interpersonal interactions of several kinds—with peers, in study groups, and with faculty. Hispanic students appear to participate in MEP programs when their grades are low, when African American participation occurs more as a routine matter.

Fourth, matriculation in Minority-Serving Institutions affects each of these student groups differently. For Black students, attending an HBCU was associated with an important graduation advantage. For Hispanic students, no such advantage accrued to them from matriculation in an HSI. This kind of difference in institutional impact has been found before by Bridges et al. (2008), and noted by Contreras et al. (2008). This finding also agrees with others that attending an HSI had no independent effect on whether underrepresented minority students completed STEM degrees (Figueroa, Hurtado, & Eagan, 2013; Nunez & Elizondo, 2015). This study confirms these trends for engineering students. Note that while UT El Paso was a high performing HSI in Chapter 3, this school did not participate in the online survey.

Fifth, by gender, there were few differences within each ethnic group, but somewhat more among Hispanic students. Among African Americans, the major difference was that Black female students were more Global (i.e., theoretical) than males. There was no indication of what might help account for their somewhat better overall GPA despite the same test scores as Black male students. Among Hispanic students, female students appear to be more outgoing. Success for them depends more on interpersonal interactions with peers and faculty. Success for Hispanic males depends more on ability and academic skills. Hispanic women have higher test scores, but achieve the same grades as males, with ability controlled.

It has been suggested that Black and Hispanic engineering students have less access to certain forms of social capital, but substitutions may occur that facilitate their longevity in the pipeline (Dika et al., 2016). Hispanic students reported less positive interactions with engineering faculty by the senior year, compared to freshmen. This finding is worrisome, since faculty are so often cited as critical in the success of engineering students (Dika, 2012; Hackett et al., 1992). Indeed, it is the general failure to nurture minority talent and its consequences that have fueled the studies of minority student adjustment. This general failure is also the impetus behind the establishment of the Minorities in Engineering Programs now supported by universities and corporations alike. Hispanic students were less involved in MEP programming unless their grades were low. Yet Hispanic students, especially women, were outgoing, and able to establish a wide range of interpersonal and social relations that facilitate academic success. Although Hispanic-Serving Institutions may not carry the graduation advantage that is true of HBCUs for Black students, Hispanic students may not suffer from the same level of avoidance as Black students.

For African American students, it certainly appears that their greater immersion in these programs provides significant interactions with peers, staff, faculty, and industry co-workers that facilitate their academic success. This MEP immersion in PWIs offers a substitute for the relative neglect that Black students could experience. On the other hand, there was clear evidence that matriculation in one of the three HBCUs in this study offers an alternative to any lack of institutional support at White schools.

References

Allen, W. R. (1992). The color of success: African American college student outcomes at predominantly white and historically black public colleges and universities. *Harvard Educational Review*, 62(1), 26–44.

Amelink, M., & Meszaros, P. (2011). A comparison of educational factors promoting or discouraging the intent to remain in engineering by gender. *European Journal of Engineering Education*, 36(1), 47–62.

Ancis, J. R., Sedlacek, W. E., & Mohr, J. J. (2000). Student perceptions of campus climate by race. *Journal of Counseling and Development*, 78(2), 180–186.

Barlow, A. E. L., & Villarejo, M. (2004). Making a difference for minorities: Evaluation of an educational enrichment program. *Journal of Research in Science Teaching*, 41(9), 861–881.

Bonner, F. A. (2014). Academically gifted African American males: Modeling achievement in the historically black colleges and universities and predominantly white institutions context. In F. A. Bonner (Ed.), *Building on resilience:*

Models and frameworks of black male success across the P-20 pipeline (pp. 109–124). Sterling, VA: Stylus Publishing.

Bridges, B. K., Kinzie, J., Laird, T. F. N., & Kuh, G. D. (2008). Student engagement and student success at historically Black and Hispanic-serving institutions. In M. Gasman, B. Baez, & C. S. V. Turner (Eds.), *Understanding minority-serving institutions* (pp. 217–236). Albany, NY: State University of New York Press.

Brown, A. R., Morning, C., & Watkins, C. (2005). Influence of African American engineering student perceptions of campus climate on graduation rates. *Journal of Engineering Education*, 94(4), 263–271.

Chiang, L., Hunter, C. D., & Yeh, C. J. (2004). Coping attitudes, sources, and practices among Black and Latino college students. *Adolescence*, 39(156), 793–815.

Chubin, D. E., Donaldson, K., Olds, B., & Fleming, L. (2008). Educating generation net: Can U.S. engineering woo and win the competition for talent? *Journal of Engineering Education*, 97(3), 245–258.

Cole, D. (2008). Constructive criticism: The role of student-faculty interactions on African American and Hispanic students' educational gains. *Journal of College Student Development*, 49(6), 587–605. doi: 10.1353/csd.0.0040.

Cole, D., & Espinoza, A. (2008). Examining the academic success of Latino students in science technology engineering and mathematics (STEM) majors. *Journal of College Student Development*, 49(4), 285–300. doi: 10.1353/csd.0.0018.

Conrad, C., & Gasman, M. (2015). *Educating a diverse nation: Lessons from minority-serving institutions*. Cambridge, MA: Harvard University Press.

Contreras, F. E., Malcolm, L. E., & Bensimon, E. M. (2008). Hispanic-serving institutions: Closeted identity and the production of equitable outcomes for Latino/a students. In M. Gasman, B. Baez, & C. S. V. Turner (Eds.), *Understanding minority-serving institutions* (pp. 71–90). Albany, NY: State University of New York Press.

Constantine, J. M. (1995). The effect of attending historically black colleges and universities on future wages of black students. *Industrial & Labor Relations Review*, 48(3), 531–546.

Constantine, M. G., Chen, E. C., & Ceesay, P. (1997). Intake concerns of racial and ethnic minority students at a university counseling center: Implications for developmental programming and outreach. *Journal of Multicultural Counseling and Development*, 25(3), 210–218.

Constantine, M. G., Wilton, L., & Caldwell, L. D. (2003). The role of social support in moderating the relationship between psychological distress and willingness to seek psychological help among Black and Latino college students. *Journal of College Counseling*, 6(2), 155–165.

Dika, S. L. (2012). Relations with faculty as social capital for college students: Evidence from Puerto Rico. *Journal of College Student Development*, 53(4), 596–610.

Dika, S. L., Pando, M. A., & Tempest, B. (2016). *Investigating the role of interaction, attitudes, and intentions for enrollment and persistence in engineering among underrepresented minority students*. Proceedings of the 2016 Annual Conference of the American Society of Engineering Education (ASEE), New Orleans, LA. Retrieved from https://peer.asee.org/17069.

Felder, R. M., Felder, G. N., & Dietz, E. J. (2002). The effects of personality type on engineering student performance and attitudes. *Journal of Engineering Education*, 91(1), 3–17.

Felder, R. M., & Silverman, L. K. (1988). Learning and teaching styles in engineering education. *Engineering Education, 78*(7), 674–681.

Figueroa, T., Hurtado, S., & Eagan, K. (2013). *Making it! . . . or not: Institutional contexts and biomedical degree attainment.* Paper presented at the meeting of the Association for Institutional Research, Long Beach, CA.

Fleming, J. (1984). *Blacks in college.* San Francisco: Jossey-Bass.

Fleming, J. (2001). The impact of an historically black college on African American students: The case of Lemoyne-Owen college. *Urban Education, 36*(5), 587–610.

Fleming, J. (2012). *Enhancing the academic performance and retention of minorities: What we can learn from program evaluation.* San Francisco: Jossey-Bass.

Fleming, J., Garcia, N., & Morning, C. (1996). The critical thinking skills of minority engineering students: An exploratory study. *Journal of Negro Education, 64*(4), 437–453.

Fleming, J., & Morning, C. (1998). Correlates of the SAT in minority engineering students: An exploratory study. *Journal of Higher Education, 69*(1), 89–108.

Flowers, A. M. (2014). Gifted, Black, male, and poor in science, technology, engineering, and mathematics. In F. A. Bonner (Ed.), *Building on resilience: Models and frameworks of black male success across the P-20 pipeline* (pp. 124–139). Sterling, VA: Stylus Publishing.

Gasman, M. (2008). Minority-serving institutions: A historical backdrop. In M. Gasman, B. Baez, & C. S. V. Turner (Eds.), *Understanding minority-serving institutions* (pp. 18–27). Albany, NY: State University of New York Press.

Hackett, G., Betz, N. E., Casas, J. M., & Rocha-Singh, I. A. (1992). Gender, ethnicity, and social cognitive factors predicting the academic achievement of students in engineering. *Journal of Counseling Psychology, 39*(4), 527.

Hacker, A. (2010). *Two nations: Black and White, separate, hostile, unequal.* New York: Simon and Schuster.

Hardy, F. R. (1974). Social origins of American scientists and scholars. *Science, 185*(4150), 497–586.

Hurtado, S., Eagan, M. K., Tran, M., Newman, C., Chang, M. J., & Velasco, P. (2011). "We do science here": Underrepresented students' interactions with faculty in different college contexts. *Journal of Social Issues, 67*(3), 553–579.

Kim, M. M. (2002). Historically black vs. white institutions: Academic development among black students. *Review of Higher Education, 25*(4), 385–407.

Kim, M. M., & Conrad, C. F. (2006). The impact of historically black colleges and universities on the academic success of African American students. *Research in Higher Education, 47*(4), 399–427.

Kim, Y. K., & Sax, L. J. (2009). Student-faculty interaction in research universities: Differences by student gender, race, social class, and first- generation status. *Research in Higher Education, 50*(5), 437–459.

Landis, R. (2005). *Retention by design.* New York: National Action Council for Minorities in Engineering.

Lent, R. W., Brown, S. D., Sheu, H. B., Schmidt, J., Brenner, B. R., Gloster, C. S., & Treistman, D. (2005). Social cognitive predictors of academic interests and goals in engineering: Utility for women and students at Historically Black Universities. *Journal of Counseling Psychology, 52*(1), 84–92.

Lewis, B. F. (2003). A critique of the literature on the underrepresentation of African Americans in science: Directions for future research. *Journal of Women and Minorities in Science and Engineering, 9*(3/4), 361–373.

Malcolm-Piquex, L. E., & Malcolm, S. (2015). African American women and men into engineering: Are some pathways smoother than others? In J. B. Slaughter, Y. Tao, & W. Pearson, Jr. (Eds.), *Changing the face of engineering: The African American experience* (pp. 90–119). Baltimore, MD: Johns Hopkins University Press.

Martin, J. P., Simmons, D. R., & Yu, S. L. (2013). The role of social capital in the experiences of Hispanic women engineering majors. *Journal of Engineering Education, 102*(2), 227–243.

Maton, K. I., & Hrabowski III, F. A. (2004). Increasing the number of African American PhDs in the sciences and engineering: A strengths-based approach. *American Psychologist, 59*(6), 547–556.

McPherson, E. (2017). Oh you are smart: Young, gifted African American women in STEM majors. *Journal of Women and Minorities in Science and Engineering, 23*(1), 1–14.

Moore, J. L. (2006). A qualitative investigation of African American males' career trajectory in engineering: Implications for teachers, school counselors, and parents. *Teachers College Record, 108*(2), 246–266.

Moore, J. L., Madison-Colmore, O., & Smith, D. M. (2003). The prove-them-wrong syndrome: Voices from unheard African-American males in engineering disciplines. *The Journal of Men's Studies, 12*(1), 61–73.

Moos, R. H. (1979). *Evaluating educational environments: Procedures, measures, finding and policy implications.* San Francisco: Jossey-Bass.

Morning, C., & Fleming, J. (1994). Project preserve: A program to retain minorities in engineering. *Journal of Engineering Education, 83*(3), 237–242.

NACME (National Action Council for Minorities in Engineering). (2014a). African Americans in engineering. *Research & Policy Brief, 4*(1, April), 1–2.

NACME (National Action Council for Minorities in Engineering). (2014b). Latinos in engineering. *Research & Policy Brief, 4*(3, October), 1–2.

Nasim, A., Roberts, A., Harrell, J. P., & Young, H. (2005). Non-cognitive predictors of academic achievement for African Americans across cultural contexts. *Journal of Negro Education, 74*(4), 344–348.

Nixon, C. T., & Frost, A. G. (1990). The study habits and attitudes inventory and its implications for students' success. *Psychological Reports, 66*(3c), 1075–1085.

Nunez, A.-M., & Elizondo, D. (2015). Institutional diversity among four-year Hispanic-serving institutions. In A. M. Nunez, S. Hurtado, & E. C. Galdeano (Eds.), *Hispanic-serving institutions: Advancing research and transformative practice* (pp. 65–81). New York: Routledge.

Oliver, M. L., Rodriguez, C. J., & Mickelson, R. A. (1985). Brown and black in white: The social adjustment and academic performance of Chicano and Black students in a predominately White university. *The Urban Review, 17*(1), 3–24.

Pearson, W. (2005). *Beyond small numbers: Voices of African American PhD chemists.* New York: Elsevier Science.

Person, D. R., & Fleming, J. (2012). Who will do math, science, engineering and technology? Academic achievement among minority students in seventeen institutions. In J. Fleming, *Enhancing the retention and academic performance of minorities: What we can learn from program evaluation* (pp. 120–134). San Francisco: Jossey-Bass.

Ramseur, H. (1975). *Continuity and change in Black identity: A study of Black students at an interracial college.* Unpublished doctoral dissertation, Cambridge MA: Harvard University.

Ransom, T. (2015). Clarifying the contributions of historically black colleges and universities in engineering education. In J. B. Slaughter, Y. Tao, & W. Pearson, Jr. (Eds.), *Changing the face of engineering: The African American experience* (pp. 120–148). Baltimore, MD: Johns Hopkins University Press.

Reichert, M., & Absher, M. (1997). Taking another look at educating African American engineers: The importance of undergraduate retention. *Journal of Engineering Education*, 86(3), 241–253.

Sedlacek, W. E. (1998). Admissions in higher education: Measuring cognitive and non-cognitive variables. In D. J. Wilds & R. Wilson (Eds.), *Minorities in higher education 1997–98* (pp. 47–68). Washington, DC: American Council on Education.

Sedlacek, W. E. (2004). *Beyond the big test: Noncognitive assessment in higher education*. San Francisco: Jossey-Bass.

Solberg, V. S., & Viliarreal, P. (1997). Examination of self-efficacy, social support, and stress as predictors of psychological and physical distress among Hispanic college students. *Hispanic Journal of Behavioral Sciences*, 19(2), 182–201.

Suarez-Balcazar, Y., Orellana-Damacela, L., Portillo, N., Rowan, J. M., & Andrews-Guillen, C. (2003). Experiences of differential treatment among college students of color. *The Journal of Higher Education*, 74(4), 428–444.

Tomlinson-Clarke, S. (1998). Dimensions of adjustment among college women. *Journal of College Student Development*, 39(4), 364–372.

Vogt, C. M. (2008). Faculty as a critical juncture in student retention and performance in engineering programs. *Journal of Engineering Education*, 97(1), 27–36.

Wolfe, J., Powell, B. A., Schlisserman, S., & Kirshon, A. (2016). *Teamwork in engineering undergraduate classes: What problems do students experience?* Proceedings of the 123rd Annual Meeting of the American Association of Engineering Education (ASEE), New Orleans, LA. ID Number 16447.

Yoder, B. (2012). *Going the distance in engineering education: Best practices and strategies for retaining engineering, engineering technology, and computing students*. Washington, DC: American Association for Engineering Education (ASEE).

8 The Intersection of Gender and Ethnicity in the College Adjustment of Engineering Students

Summary

Despite numerous studies of lack of belonging issues for women in engineering education, and reportedly more gender issues than race issues, the women in this survey performed as well or better than male counterparts, and their seemingly better social skills contribute to their success rather than setting them apart. For all students, participation in academic programs was the key to success. This pattern was truer of women than men, and most true of minorities. Access to the success factors of internships and research may be the key issues, since access to them clearly favors White students.

Introduction

After completing their online surveys, several women did complain that although there were questions addressing prejudice and racism, there were no questions that directly addressed sexism. At the time, we were confident that any pattern of gender differences would shed light on sexism, as well as other gender issues. But so far in these analyses, gender has not emerged as a startling issue—and certainly not as an encumbrance for women. That being the case, it seemed that a full dissection of gender within and across ethnicity was warranted before the discussion was closed. A brief review of studies on gender in engineering provides a context in which to place the results.

Gender issues are reportedly significant in engineering due to women being vastly outnumbered. Males constitute approximately 80% of the undergraduate engineering population, and White males are still the dominant constituents (NCES, 2013). This disparity is unusual today, and only exists in a few areas including computer science and physics, while other fields have become gender balanced. So vast is the difference in their numbers in engineering that gender issues have been reported to be more problematic than racial issues (Van Aken, Watford, & Medina-Borja, 1999; see Chapter 4), although a web survey found that Black women saw race as more of a barrier than gender and saw gender as more of a barrier than did White women (Hanson, 2006). Cheryan,

Ziegler, Montoya, and Jiang (2017) sum up the reasons for gender inequity in engineering and several other still-almost-all-male majors as due to: a masculine environment that raises belonging issues; women's lack of historical experience with engineering; and gaps in self-efficacy. The research does seem to support their conclusions.

Women experience the engineering climate differently and appear to suffer from a sense of not belonging, and a belief that they receive different treatment than men (Heyman, Martyna, & Bhatia, 2002). While the chilly climate is not exclusive to engineering, it is shaped by the attitudes and behaviors of male colleagues who create a less than comfortable place for women (Ferreira, 2002; Mills, Bastalich, Franzway, Gill, & Sharp, 2006). The underrepresentation of women in engineering has been attributed to "social" barriers in the classroom (Vogt, 2003), the experience of negative interactions that leads women to become interested in other majors (Litzler & Young, 2012), the greater experience of incivility (Miner, Diaz, & Rinn, 2017), as well as the experience of discrimination (Ayre, Mills, & Gill, 2011). According to Allendoerfer et al. (2012), providing opportunities for belonging gives the most return on the persistence investment, but women report less sense of it (Metz, Brainard, & Gillmore, 1999). Indeed, Marra, Rodgers, and Shen (2012) found lack of belonging to be the strongest factor in leaving engineering in general, and the strongest factor for minorities, while Fifolt and Abbott (2008) reported less psychosocial and career-related support for women in engineering. Teamwork problems have been the subject of frequent complaints, including forcing women into the traditional roles of secretary or worse, being ignored (Camacho & Lord, 2013). In the focus groups of Chapter 4, women also reported more teamwork problems, being relegated to secretarial roles in groups, having their comments devalued, and being ignored, while racial issues were very much understated. In Camacho and Lord's (2013) focus groups, women reported that male students were listened to, while females saying the same thing were dismissed. In addition, the engineering environment is so competitive that friends are hard to come by, although Camacho and Lord (2013) found that White females were more likely to have friends than minority women. Espinosa (2008) found that the college environment was more important than background factors or pre-college in encouraging women to stay in science. What works to retain women are friendly and informal relationships between faculty members and students, according to Whitten et al. (2004), who took lessons from visits to HBCUs, which are extraordinarily productive of female scientists. Perna et al. (2009) also reported that one HBCU, Spelman College, was successful in retaining women in STEM because of low levels of social isolation, alienation, and racism, as well as creating a more supportive social, cultural, and racial environment. Miner et al. (2017) concur that social support alleviates the experience of distress due to incivility. For Hispanic women, peers and institutional support programs can alleviate a lack of social capital (Martin, Simmons, & Yu, 2013). Because students of color experience more incivility than women (e.g., Miner et al., 2017), we infer that women of color experience a magnified sense of negative interactions. Indeed, Ong, Wright, Espinosa, and Orfield (2011) emphasized the critical role of climate for women of color in STEM, including the experiences of isolation, identity, invisibility, lack of belonging, micro-aggressions, and difficulty navigating the academic environment; they also point out the impact of positive words and academic programs. For women of color transferring from community colleges, the treatment issues are compounded by age and the community college stigma (Reyes, 2011). Tate and Linn (2005) stressed the multiple identities that women of color in engineering grapple with, such that academic success is anchoring but race is salient in their social arenas.

Even women who survive to become engineering faculty perceive their departments as less tolerant of diversity than do male counterparts (Billmoria, Joy, & Liang, 2008). Female engineers in Australia described the workplace as a boy's club that engendered lack of opportunity, lack of support, and difficult coping strategies that contributed to feelings of work discomfort (Mills et al., 2006). Fewer women than men described their science workplaces in positive terms, and more often described them as uncomfortable, with frequent tense or hostile interactions (Gunter & Stambach, 2005). In short, engineering education and the engineering workplace is less comfortable for women.

While all of this may be true, Vogt (2003, p. 217) described a "new turn of events" where women entering engineering may have higher grades and a greater tendency to remain than men. The reason may be that women apply more effort to combat their lower feelings of self-efficacy. Kimball, Cole, Hobson, Watson, and Stanley (2008) found that women in engineering at Texas A&M completed required core "barrier" courses faster than males, while African American and Hispanic women completed these courses faster than males of the same ethnicity. These findings held across five engineering majors studied. Camacho and Lord (2013) reported exceptional performance and retention among Latinas, perhaps the most marginalized group in engineering. Latinas were also the most successful of all transfer students, with a 72% persistence rate (Camacho & Lord, 2013), while Varma (2002) found Hispanic women to be the most independent and assertive in getting their academic needs met.

In several studies, women in engineering reported lower math self-concept (Miner et al., 2017), lower expectancies of success (Lee, Brozina, Amelink, & Jones, 2017), and less confidence in their abilities (Heyman et al., 2002; Metz et al., 1999; Wilson, Bates, Scott, Painter, & Shaffer, 2015), even after controls for critical factors (Litzler, Samuelson, & Lorah, 2014), less self-efficacy (Vogt, 2003), and less experience with mechanical things (Micari, Pazos, & Hartmann, 2007). Women also expect more barriers to their engineering career attainment (Infante-Perea, Román-Onsalo, & Navarro-Astor, 2018), and expressed lower master's degree aspirations than men, although minorities expressed higher master's level aspirations (Litzler & Lorah, 2018). Despite less confidence, they performed as well as men (Micari et al., 2007; Varma, 2002). The compensating effort notion gains credence with findings that Asian students reported the lowest levels of self-efficacy in their study, but not lower performance (Wilson et al., 2015). For women of color in engineering, Tate and Linn (2005) reported they may be plagued by feelings of difference and not belonging with uncomfortable social interactions, but that academic success is anchoring and that they are fully engaged in coursework and on target to complete their programs.

Certainly, in our studies so far, the performance of women seems to suffer less than their feelings of belonging. The study of institutional statistics found that across ethnicity, women performed better than males in the classroom (see Chapter 3). Although we were not able to disaggregate non-minority populations by ethnicity, non-minority women achieved higher average GPAs than non-minority men, especially after a control for test scores. They also had higher math grades, and higher 6-year graduation rates than non-minority males (57.6% vs. 47.4%). Among minorities, Black women had higher overall GPA than Black men, as well as higher GPA in math and engineering core courses, but did not exhibit higher retention rates. Hispanic women achieved higher overall GPA, but not higher graduation rates compared to Hispanic males. Thus, better academic performance yielded better graduation rates only for non-minority females, even with test scores controlled. This race by gender effect begs for an explanation. In previous survey analyses of Chapters 6–7, there were few gender differences on study measures or in resulting success factors unearthed. What the gender differences did suggest was better social adjustment among women, which may mean better social skills.

Conclusion

From the foregoing, we can conclude that due to the gender inequities in numbers, women find the engineering educational environment to be uncomfortable in a number of respects, culminating in lower feelings of self-confidence. Women of color appear to share in the sense of discomfort or lack of belonging, although HBCUs offer a relative haven from the chilly climate. Despite any discomfort women may experience, their performance does not seem to suffer, and they may even outperform their male counterparts.

Purpose of the Study

Given the relatively minor impact of gender—or rather, the positive rather than negative impact of gender so far—what more light can we shed on the intersection of ethnicity and gender in the survey study of program participation, adjustment, learning styles and their consequences for successful performance? We can look at things a bit differently in this set of analyses. The survey data affords us an opportunity to dissect gender issues for non-minorities and minorities, but also to look at gender within five ethnic groups—White, Asian, Middle Eastern, Black, and Hispanic. Previous analyses in this study have only examined non-minorities as a whole, so this chapter provides a rare glimpse into all groups represented in the survey. Not only are gender differences within ethnicity examined, but ethnic differences were examined within gender.

Then we examine gender/ethnic differences in study measures associated with three measures of success—GPA, effective school rankings, and freshman-senior differences or retention—using regression procedures. The challenge was to find succinct ways of describing these differences. Previous chapters utilized factor analytic techniques, while this chapter resorts to descriptive statistics for three reasons. They provide a more succinct method of describing gender and ethnic differences; they allow for a weighting of multiple predicting factors; and they avoid the small sample size problems of factor analysis that would raise validity questions for the race and gender intersections. In all analyses, there were methods of either controlling for differences in test scores, as in analysis of covariance for mean differences in study measures, or in determining the contributions of tested abilities in predicting success on three measures, as in regression procedures.

Method

Study Population

The survey population table was presented in Chapter 6. There were a total of 1,186 students, 37.6% of whom were female. Minorities comprised 632 (or 55.2%) of the sample. There were 225 (or 35.6%) African American, 35 (or 5.5%) African-Caribbean (for a total of 260 Black students), and 368 (or 58.2%) were Hispanic. There were 41 biracial students and four Native American students who were represented only in whole-sample analyses; they were not represented in ethnic breakdowns. Female students comprised 39.7% of the minority group. There was a higher percentage of females among African Americans than among Hispanics (45.3% vs. 34.5%). When minority students were broken down into ethnic groups, only African American and Hispanic students were included. Of the 513 non-minority students, 35.1% were female. White students constituted 56.7% of the group, with 35.1% female; Asian students were 26.9%, with 36.9% female; and Middle Eastern students were 16.4% of the sample, with 32.1% female. Note that the percentage of female students does not differ substantially among non-minority groups, that the proportion of female students was highest among African

Americans, and that the female percentage in this survey is substantially higher than the 20% in general in engineering.

Measures

An online survey including 11 measures of program participation and college adjustment were administered via Survey Monkey to students attending 18 institutions. This analysis compared gender differences in the study measures within each of the five ethnic groups, and within each group determined the most important variables predicting three measures of "success:" overall GPA; matriculation in engineering schools more effective in graduating minorities; and freshman-senior differences suggesting better retention.

The measures, which included program involvement, college adjustment, and learning styles, were adapted from the evaluation of nine different programs in various STEM fields, as well as studies of student development in college (Felder & Silverman, 1988; Felder, Felder, & Dietz, 2002; Fleming, 1984, 2001, 2012; Hardy, 1974; Moos, 1979; Nasim, Roberts, Harrell, & Young, 2005; Nixon & Frost, 1990; Pearson, 2005; Person & Fleming, 2012; Ramseur, 1975; Sedlacek, 1998, 2004). Measures of program participation include an overall scale of general program participation in 11 program components, as well as separate assessments of participation in each of the 11 programs. Measures of college adjustment included those specific to college adjustment dimensions—College Adjustment, Study Habits, Academic Adjustment, Social Adjustment, Faculty Interactions, and Comfort, as well as those measuring Community Identity, Scientific Orientation, and Learning Styles. Throughout the chapter they are collectively referred to as measures of program participation, college adjustment, and Learning Styles. The details of instrumentation and analysis are presented in Appendices 1–3 of Chapter 6.

Gender Differences in Study Measures

The questions to keep in mind are: what are the nature of gender difference within each of the five ethnic groups; are the gender differences within minorities different from those within non-minorities; and what are the probable consequences of any gender differences for women?

Gender Differences Across Ethnicity

If we look at all men and women in the entire survey study, the analyses showed no disadvantages for women (Table 8.1). Female grades were slightly better with test scores controlled. Women participated more than men in programs. This includes significant but small differences in favor of women on High School STEM Programs, Tutoring, Supplemental Instruction, Study

Table 8.1 Gender Differences in Program Participation and College Adjustment Measures

All Students (n = 1,186)	Male Mean n = 740	Female Mean n = 446	F with Test Score Percentile	Effect Size (Eta)
DEMOGRAPHIC VARIABLES:				
GPA	3.08 (.019)	3.14 (.025)	4.12*	.003
GPA, With Test Score Percentile	3.08 (.019)	3.15 (.024)	3.13 $p < .10$	.003
Test Score Percentile	77.22 (.742)	78.82 (.956)	ns	na
PROGRAM PARTICIPATION SCALES:				
Program Participation Scale	27.4 (.348)	29.3 (.448)	11.43***	.010
High School STEM Program Participation Scale	2.69 (.085)	3.14 (.110)	10.45***	.009
Tutoring Participation Scale	3.14 (.087)	3.53 (.112)	16.39***	.014
Supplemental Instruction Participation Scale	2.99 (.086)	3.29 (.111)	9.07**	.008
Study Group Participation Scale	2.83 (.087)	3.11 (.113)	3.91~*	.004
Peer Mentoring Participation Scale	1.85 (.067)	2.09 (.087)	5.21	.004
COLLEGE ADJUSTMENT SCALES:				
Academic Adjustment Scale	57.8 (.308)	59.3 (.397)	9.18***	.008
Engineering Faculty Interactions Scale	118.0 (.740)	120.8 (.954)	5.09*	.004
Academic Management Scale	116.4 (.761)	120.9 (.981)	13.08***	.011
Study Habits Scale	64.1 (.522)	66.4 (.672)	7.39**	.006
Social Adjustment Scale	100.8 (.684)	106.9 (.881)	30.29***	.025
Community Identity Scale	96.6 (.520)	99.7 (.669)	13.69***	.011
LEARNING STYLES:				
Visual	8.08 (.078)	7.62 (.100)	13.36***	.011
Auditory	2.92 (.078)	3.38 (.100)	13.36***	.011

(*Continued*)

Table 8.1 (Continued)

White Students (n = 291)	White Male Mean n = 189	White Female Mean n = 102	F with Test Score Percentile	Effect Size (Eta)
DEMOGRAPHIC VARIABLES:				
GPA	3.23 (.031)	3.23 (.050)	ns	na
GPA, With Test Score Percentile	3.23 (.036)	3.23 (.049)	ns	na
Test Score Percentile	86.1 (.907)	85.4 (1.24)	ns	na
PROGRAM PARTICIPATION SCALES:				
Summer Bridge Participation Scale	1.25 (.091)	1.58 (.124)	4.51*	.015
Supplemental Instruction Participation Scale	2.88 (.164)	3.52 (.224)	5.31*	.018
COLLEGE ADJUSTMENT SCALES:				
Social Adjustment Scale	102.11 (1.46)	109.42 (1.99)	8.76**	.030
LEARNING STYLES:				
Visual	8.25 (.158)	7.31 (.215)	12.45***	.041
Auditory	2.75 (.158)	3.69 (.215)	12.45***	.041

Asian Students (n = 138)	Asian Male Mean n = 87	Asian Female Mean n = 51	F with Test Score Percentile	Effect Size (Eta)
DEMOGRAPHIC VARIABLES:				
GPA	3.06 (.057)	3.41 (.075)	13.37***	.089
GPA, With Test Score Percentile	3.07 (.057)	3.39 (.074)	11.94***	.081
Test Score Percentile	81.1 2.87)	85.5 (2.86)	ns	na
PROGRAM PARTICIPATION SCALES:	No Results			
COLLEGE ADJUSTMENT SCALES:				
College Adjustment Scale	66.9 (1.64)	72.3 (2.15)	3.94*	.028
Academic Adjustment Scale	55.1 (.953)	58.9 (1.25)	6.14*	.044
Social Adjustment Scale	96.7 (1.79)	105.3 (2.35)	8.54**	.059
LEARNING STYLES:	No Results			

(*Continued*)

Table 8.1 (Continued)

Middle Eastern Students (n = 84)	Middle Eastern Male Mean n = 57	Middle Eastern Female Mean n = 27	F with Test Score Percentile	Effect Size (Eta)
DEMOGRAPHIC VARIABLES:				
GPA	3.09 (.074)	3.11 (.108)	ns	na
GPA, With Test Score Percentile	3.09 (.074)	3.11 (.108)	ns	na
Test Score Percentile	81.2 (2.69)	80.1 (3.92)	ns	na
PROGRAM PARTICIPATION SCALES:				
High School STEM Program Participation Scale	2.37 (.305)	3.59 (.443)	5.17*	.060
COLLEGE ADJUSTMENT SCALES:	No Results			
LEARNING STYLES:	No Results			

Black Students (n = 260)	Black Male Mean n = 137	Black Female Mean n = 123	F with Test Score Percentile	Effect Size (Eta)
DEMOGRAPHIC VARIABLES:				
GPA	2.99 (.046)	2.99 (.050)	ns	na
GPA, With Test Score Percentile	2.99 (.045)	2.99 (.049)	ns	na
Test Score Percentile	74.4 (1.89)	74.6 (2.08)	ns	na
PROGRAM PARTICIPATION SCALES:	No Results			
COLLEGE ADJUSTMENT SCALES:	No Results			
LEARNING STYLES:				
Visual	8.10 (.201)	7.40 (.221)	5.49*	.024
Auditory	2.89 (.201)	3.59 (.221)	5.49*	.024
Sequential	6.66 (.182)	6.01 (.200)	5.73*	.025
Global	4.34 (.182)	4.99 (.200)	5.73*	.025

(Continued)

Table 8.1 (Continued)

Hispanic Students (n = 368)	Hispanic Male Mean n = 241	Hispanic Female Mean n = 127	F with Test Score Percentile	Effect Size (Eta)
DEMOGRAPHIC VARIABLES:				
GPA	3.02 (.032)	3.10 (.424)	ns	na
GPA, With Test Score Percentile	3.03 (.031)	3.08 (.042)	ns	na
Test Score Percentile	68.5 (1.43)	73.7 (1.97)	4.56*	.012
PROGRAM PARTICIPATION SCALES:				
Program Participation Scale	26.9 (.607)	29.3 (.838)	5.11*	.014
High School STEM Program Participation Scale	2.42 (.146)	3.11 (.202)	7.68**	.021
COLLEGE ADJUSTMENT SCALES:				
College Adjustment Scale	73.2 (.852)	76.6 (.176)	5.25*	.014
Academic Adjustment Scale	59.0 (.478)	61.2 (.660)	6.79**	.018
Engineering Faculty Interactions Scale	118.6 (1.23)	124.5 (1.70)	7.74**	.021
Academic Management Scale	117.5 (1.304)	123.7 1.80)	7.82**	.021
Social Adjustment Scale	100.8 (.761)	109.7 (1.55)	21.42***	.055
Community Identity Scale	96.7 (.874)	101.8 (1.21)	11.66***	.031
LEARNING STYLES:	No Results			

Note. Standard error of the mean in parenthesis. Effect size (partial *Eta* squared) given for significant results.
* $p < .05$
** $p < .01$
*** $p < .001$

Groups, and Peer Mentoring. There were no differences, however, in the important areas of Faculty Mentoring, Research or Internship participation. Women reported better adjustment on six of nine scales. The strongest effect was better Social Adjustment, followed by stronger Community Identity, better Academic Management and Academic Adjustment. Significant but smaller effects were found for better Faculty Interactions, and better Study Habits. Men reported stronger Visual Leaning Styles, while women reported stronger Auditory Learning Styles. Both genders reported stronger Visual than Auditory Learning Styles. In short, these differences showed no disfavor for women. Recall that the analyses controlled for test scores, so these are men and women with roughly equal tested abilities. Women, then, may complain of or bear gender bias, but they appear to work harder in ways that pay off and compensate for bias. They were more involved—in programs that have academic benefits, and in social areas that may buffer them. They even report better relationships and interactions with faculty, even though they did not receive more faculty mentoring than men.

Gender Differences Among Non-Minority Students

Gender differences among White students were few and far between (Table 8.1). There were no gender differences in GPA among White students. There were only two differences in program participation, with females participating more in Summer Bridge and Supplemental Instruction. On adjustment measures, females reported better Social Adjustment. On Learning Styles, males were more Visual, while females were more Auditory, although both genders were more Visual than Auditory. In short, there were few gender differences, with differences in Visual and Auditory Learning Styles being the strongest.

Gender differences among Asian students were the most notable, with differences in GPA and adjustment favoring Asian women. Unlike White females, Asian females achieved significantly better grades than Asian males, with or without test scores controlled, although there were no differences in test scores. The average Asian female GPA of 3.4 was stellar, while other groups hovered closer to 3.0. There were no differences in program participation. On adjustment measures, females reported better Social Adjustment, College Adjustment, and Academic Adjustment. There were no differences in Learning Styles. In short, there were more gender differences among Asian students yielding better academic performance and better adjustment for Asian women.

Gender differences among Middle Eastern students were close to non-existent. There were no differences in GPA, test scores, or GPA after test scores were controlled. The only difference in program participation was that females participated more in High School STEM Programs. There were no differences at all on adjustment measures, not even on social adjustment. There were no differences in Learning Styles. In short, there were virtually no gender differences among Middle Eastern students.

What seems most significant is that gender differences were minimal among non-minorities, except for Asian students, where there were gender differences greatly favoring Asian females. The Asian female GPA advantage was stunning, and consistent with their better adjustment. While our expectations may be high for Asian students as a whole, we were not prepared for the overperformance of Asian women more satisfied with their engineering experience.

Gender Differences Among Minority Students

As shown in Table 8.1, there were no differences of great significance among African Americans—no differences in GPA or test scores, no differences in program participation, no differences in adjustment measures. On Learning Styles, Black males were more Visual and Sequential, while Black females were more Auditory and Global.

Among Hispanic students, Hispanic females had better test scores, but there were no differences in GPA (Table 8.1). Females reported more program participation in general, but only the subscale for high school program participation was significant. Females reported better adjustment on six of nine measures the strongest being Social Adjustment, followed by Community Identity, Academic Management, Faculty Interactions, Academic Adjustment, and College Adjustment. There were no gender differences in Learning Styles. The big story here is the far better adjustment of Hispanic females. The adjustment differences suggest better social skills, navigational skills, and satisfaction with engineering education. Such outcomes help explain Camacho and Lord's (2013) findings of extraordinary Latina persistence in engineering.

Ethnic Differences Within Gender

For gender analyses, it is usually sufficient compare men and women within groups, but here we look at ethnic differences among males, and then among females, and ask whether this component tells us anything that the usual male-female comparisons did not.

Ethnic Differences Among Non-Minority Males

As shown in Table 8.2, White males possess the advantage on this set of study measures. They have the highest test scores. But with test scores controlled, there were no differences among the three groups in GPA. White males reported more program participation. This was principally true for Internship Involvement, where they participate far more than the other groups. Asian males participated more in Faculty Mentoring and Research. Perhaps most important, White males scored higher on seven of nine adjustment scales, the strongest difference being College Adjustment. Asian males scored higher only on Study Habits. White males scored higher on Active learning styles while Asian males scored higher on Reflective learning styles.

In sum, with abilities matched there were no differences in grades among non-minority males, but White males reported far better adjustment and more Internship Involvement. Middle Eastern Males were not distinguished in these analyses either positively or negatively. We know from previous analyses that minorities participated more in programs than non-minority students, but among non-minority males, White students were the most participatory.

Table 8.2 Ethnic Differences in Program Participation and College Adjustment Measures for Non-Minority Students

Non-Minority Males (n = 333)	White Male Mean n = 189	Asian Male Mean n = 87	Middle Eastern Male Mean n = 57	F with Test Score Percentile	Effect Size (Eta)
DEMOGRAPHIC VARIABLES:					
GPA	3.23 (.041)	3.06 (.060)	3.09 (.074)	3.40*	.020
GPA, With Test Score Percentile	3.22 (.040)	3.08 (.059)	3.10 (.073)	ns	ns
Test Score Percentile	86.1 (1.24)	81.1 (1.83)	81.2 (2.26)	3.44*	.020
PROGRAM PARTICIPATION SCALES:					
Program Participation Scale	27.9 (.658)	25.7 (.968)	24.5 (1.19)	3.92*	.023
Faculty Mentoring Participation Scale	1.77 (.137)	2.32 (.202)	2.20 (.249)	4.06*	.024
Research Participation Scale	1.57 (.123)	2.14 (.181)	1.53 (.223)	3.79*	.023
Internship Participation Scale	4.22 (.150)	1.54 (.221)	1.88 (.273)	62.4***	.275
COLLEGE ADJUSTMENT SCALES:					
College Adjustment Scale	74.1 (1.13)	67.0 (1.66)	64.2 (2.04)	12.01***	.068
Comfort Scale	44.9 (.644)	41.1 (.949)	40.4 (1.17)	8.65***	.050
Academic Adjustment Scale	58.5 (.654)	55.1 (.964)	53.2 (1.19)	9.57***	.055
Engineering Faculty Interactions Scale	121.3 (1.41)	115.7 (2.08)	111.0 (2.57)	7.01***	.041
Academic Management Scale	118.5 (1.49)	112.0 (2.19)	108.9 (2.70)	6.12**	.036
Study Habits Scale	60.3 (1.07)	65.3 (2.19)	62.1 (1.95)	3.27*	.020
Social Adjustment Scale	102.1 (1.32)	96.7 (1.94)	96.7 (2.39)	3.79*	.023
Scientific Orientation Scale	68.9 (0.86)	63.7 (1.26)	63.01 (1.55)	8.77***	.051
LEARNING STYLES:					
Active	6.54 (.157)	5.70 (.231)	5.93 (.284)	5.06**	.030
Reflective	4.46 (.157)	5.29 (.231)	5.07 (.284)	5.06**	.030

(*Continued*)

Table 8.2 (Continued)

Non-Minority Females (n = 180)	White Female Mean n = 102	Asian Female Mean n = 51	Middle Eastern Female Mean n = 27	F with Test Score Percentile	Effect Size (Eta)
DEMOGRAPHIC VARIABLES:					
GPA	3.23 (.043)	3.41 (.061)	3.11 (.084)	4.84**	.052
GPA, With Test Score Percentile	3.22 (.043)	3.40 (.061)	3.12 (.085)	4.42*	.048
Test Score Percentile	85.4 (1.48)	85.5 (2.09)	80.1 (2.88)	ns	na
PROGRAM PARTICIPATION SCALES:					
Tutoring Participation Scale	3.00 (.229)	2.88 (.324)	2.62 (.448)	4.37*	.047
Research Participation Scale	1.59 (.179)	2.45 (.252)	1.81 (.349)	3.92*	.017
Internship Participation Scale	3.73 (.216)	1.57 (.305)	2.12 (.422)	18.76***	.176
COLLEGE ADJUSTMENT SCALES:					
College Adjustment Scale	73.2 (1.58)	72.1 (2.34)	64.6 (3.09)	3.13*	.034
Social Adjustment Scale	109.5 (2.05)	105.1 (2.90)	97.9 (4.02)	3.44*	.038
LEARNING STYLES:					
Visual	7.33 (.211)	8.51 (.298)	7.93 (.113)	5.33**	.057
Auditory	3.67 (.211)	2.49 (.298)	3.07 (.413)	5.33**	.057

Note. Standard error of the mean in parenthesis. Effect size (partial *Eta* squared) given for significant results.
* $p < .05$
** $p < .01$
*** $p < .001$

Ethnic Differences Among Non-Minority Females

Asian females outperformed other females in GPA (as well as Asian males), with or without controlling test scores, even though there were no differences in test scores. White females participated more in Tutoring and especially in Internships. Asian females participated more in

Research. White females reported the highest scores on College Adjustment and Social Adjustment. Asian female Learning Styles were the most strongly Visual, while White female Learning Styles were most strongly Auditory. All females were stronger in Visual than Auditory Learning Styles. Asian females, then, achieved the highest GPA, participated more in research, and had the highest visual learning style profiles. White females reported better adjustment and participate more in internships. Middle Eastern women were not distinguished on these measures. Despite no advantages in adjustment compared to other non-minority female counterparts, Asian females overperform academically setting an extraordinarily high bar. It is true that these 40 females come from different universities, so that environmental context is not equalized, but their grades are official and the outcome thought-provoking.

Ethnic Differences Among Minority Males

As found in Chapter 7, Black and Hispanic males appeared more similar than different (Table 8.3). Black male test scores were somewhat higher, but there were no differences in grades with or without controls for test scores. Black men participated more in program activities including Tutoring, Supplemental Instruction, and Internships. On adjustment measures, Hispanic men reported a stronger Scientific Orientation. There were no differences in Learning Styles.

Ethnic Differences Among Minority Females

Black and Hispanic women also appeared more similar than different, and the pattern of differences was virtually the same as those for Black and Hispanic men. There were no differences in test scores or grades. Black women participated more in program activities including Tutoring, Supplemental Instruction, and Internships. On adjustment measures, Hispanic women reported a stronger scientific orientation. There were no differences in learning styles.

In sum, White students reported the best adjustment, compared to their gender counterparts, but White males showed the best adjustment by far. White students had higher test scores, but as long as test scores were controlled, White students did not achieve better grades. This set of analyses was distinguished by an absence of gender difference in academic performance, with one exception: Asian females outperform all others. For Black and Hispanic students of both genders, Black students participated more in program activities while Hispanic students professed a stronger scientific orientation.

Table 8.3 Ethnic Differences in Program Participation and College Adjustment Measures for Minority Students

Minority Males (n = 378)	Black Male Mean n = 137	Hispanic Male Mean n = 241	F with Test Score Percentile	Effect Size (Eta)
DEMOGRAPHIC VARIABLES:				
GPA	2.99 (.045)	3.02 (.032)	ns	na
GPA, With Test Score Percentile	2.98 (.044)	3.03 (.032)	ns	na
Test Score Percentile	74.4 (2.05)	68.5 (1.46)	5.42*	.015
PROGRAM PARTICIPATION SCALES:				
Program Participation Scale	29.9 (.871)	26.9 (.621)	7.77**	.021
Tutoring Participation Scale	3.91 (.214)	3.19 (.153)	7.30**	.020
Supplemental Instruction Participation Scale	3.88 (.209)	2.86 (.149)	15.69***	.042
Internship Participation Scale	2.77 (.180)	1.80 (.128)	19.03***	.050
COLLEGE ADJUSTMENT SCALES:				
Scientific Orientation Scale	66.9 (1.05)	70.7 (1.04)	6.33*	.025
LEARNING STYLES:	No Results			

Minority Females (n = 250)	Black Female Mean n = 123	Hispanic Female Mean n = 127	F with Test Score Percentile	Effect Size (Eta)
DEMOGRAPHIC VARIABLES:				
GPA	2.99 (.048)	3.10 (.043)	ns	na
GPA, With Test Score Percentile	2.99 (.047)	3.11 (.042)	3.43 $p < .10$	na
Test Score Percentile	74.6 (1.99)	73.7 (1.78)	ns	na

(*Continued*)

Table 8.3 (Continued)

PROGRAM PARTICIPATION SCALES:				
Program Participation Scale	32.1 (.972)	29.1 (.871)	5.05*	.022
Tutoring Participation Scale	4.39 (.235)	3.61 (.210)	6.09*	.023
Supplemental Instruction Participation Scale	3.87 (.232)	3.15 (.208)	5.35*	.023
Internship Participation Scale	2.71 (.214)	2.03 (.192)	5.72*	.025
COLLEGE ADJUSTMENT SCALES:				
Scientific Orientation Scale	98.6 (1.54)	102.6 (1.38)	3.71*	.016
LEARNING STYLES:	No Results			

Note. Standard error of the mean in parenthesis. Effect size (partial *Eta* squared) given for significant results.
* $p < .05$
** $p < .01$
*** $p < .001$

Success Factors by Gender and Ethnicity

Now that the gender differences in the study measures have been more fully dissected, the questions are what study variables or factors contributed to success in engineering, and if they differ by gender. In previous chapters, regression equations were used to determine the predictors of GPA, matriculating in more effective engineering schools indicated by better graduation rates, and freshman-senior differences that might imply retention to the senior year(s). The predictors were combined and sorted using factor analytic techniques to describe salient common factors. In this set, however, the dissection by ethnic groups and genders increases the intersections and reduces the numbers of students to the point where factor analyses become unreliable. Furthermore, factor analysis prohibits the ability to weight factors that emerge frequently. Instead, descriptive statistics (i.e., average percentages) were used to categorize the significant predictors according to whether they represented program participation variables, adjustment variables, learning style variables, or test scores.

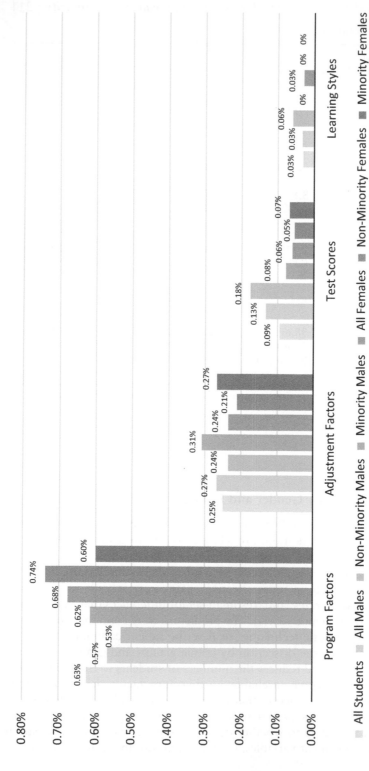

Figure 8.1 Summary of Success Factors for Five Ethnic Groups by Gender

■ All Students ■ All Males ■ Non-Minority Males ■ Minority Males ■ All Females ■ Non-Minority Females ■ Minority Females

Figure 8.1 presents a summary of success factors for all students by gender and ethnicity, while the relevant details for each ethnic group appear in Appendix 8.1. The figure displays these factors for the five ethnic groups by gender, and also presents them for the summed block of all students, as well as non-minority and minority males and females. Summed averages for five groups of all students showed that program participation measures were more important to the three success factors than adjustment measures by 2.5:1, while learning styles and test scores were far less important. Program factors accounted for 62.5% of the success factors, followed by 25.0% for adjustment factors, 9.4% for learning styles, and 3.1% for test scores. This means that greater participation in academic support program components contributed most to engineering student success.

What kinds of programs were most effective in facilitating student success? Fifty-five percent of the program factors were internships (30%) or research (25.0%), but eight programs were represented. Program participation was more important to success for women than men (67.6% vs. 56.7%), while there was little gender difference in the average importance to success of adjustment factors—23.5% for women and 26.7% for men. For women, 47.8% of the predictive program factors were internships (26.1%) or research (21.7%). For males, 64.7% were internships (35.3%) or research (29.4%). A comparison of non-minority and minority males shows that program factors were somewhat more important for minority males (61.5%) than non-minority males (52.9%). There was considerable variation within the groups of males with program factors most facilitating the success of Asian males (71.4%) among non-minorities, and Black males among minority males (80%). Among non-minority females, program factors were 73.7% of the success factors, and 60.0% for minority females. Among non-minority females program factors were most important to success for Asian females (85.7%) and White females (83.3%). Among minority females, program factors were much more important to Black female success (83.3%) than Hispanic females (44.4%), for whom the distribution of program and adjustment factors were more balanced.

Although previous analyses showed that women participated more in academic programs than males, and minorities participated more in programs than non-minorities, it appears that for all students, program participation predicted success in engineering far more than adjustment, test scores, or learning styles. While eight program components made contributions to success, internships and research participation played the strongest roles. In general, program participation was more important to success for females than males, and for minorities.

The Prediction of GPA

We can look at each of the success measures separately for greater precision. For GPA, the picture is different with adjustment factors as or

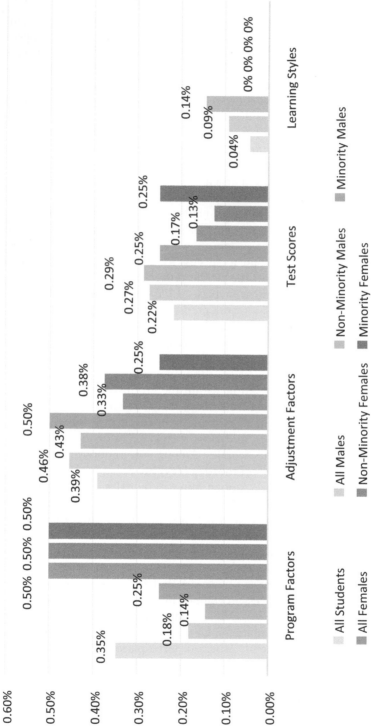

Figure 8.2 GPA Success Factors for Five Ethnic Groups by Gender

more important than program factors, but program participation factors were more important for females than males regardless of ethnic group. Figure 8.2 shows that for all students—that is, for the average of the 10 ethnic and gender groupings—adjustment factors constituted 39.1% of GPA predictors, followed by 34.8% for program factors, 21.7% for test scores, and 4.3% for learning styles. For all groups of males, adjustment factors were substantially more important in predicting GPA than program factors: adjustment—45.5%; program participation—18.2%; test scores—27.3%; and learning styles—9.1%. There was very little difference in these figures for non-minority and minority males. Which adjustment measures figured prominently in academic success for males? There were two: Academic Adjustment (i.e., satisfaction with academics in three of five cases) and Academic Management (i.e., skill in academic navigation, in two of five cases). The program factors were Research and Internships, one in each case.

For all female students, program factors were the most important predictors of GPA. Program factors constituted 50% of predictors, followed by 33.3% for adjustment actors, and 16.6% for test scores. There was no substantial difference in this pattern for non-minority and minority women. The facilitating programs were Middle School STEM Programs, Summer Bridge, Freshman Orientation Courses, Tutoring, Research, and Internships. Since no program was represented more than any other, it seems that general program involvement figured prominently in better female academic performance. The facilitating adjustment measures were Academic Management, Academic Adjustment, and Social Adjustment, with Academic Management predicting in two of four cases. Also, programs were more important for Black females, while adjustment was more important for Hispanic females.

Predictors of Effective Schools

Effective engineering schools are those with higher 6-year graduation rates after a correction for the average test scores of the student population. Chapter 3 provides a full explanation of the statistical methodology. For the present purposes, such rates were calculated for minority students, as well as separately for non-minority students. The base regression analysis tells us which variables or factors were most strongly associated with matriculation in schools with better graduation rates, while the descriptive statistics permit tabulation of results.

The key to more effective schools appears to lie solidly with participation in academic programs, and in comparison with this finding, the gender differences were minimal. Figure 8.3 shows that for all students— that is, for the average of the ten ethnic by gender groups—66.7% of the factors associated with effective schools described greater student participation in programs. College adjustment factors accounted for 22.2%,

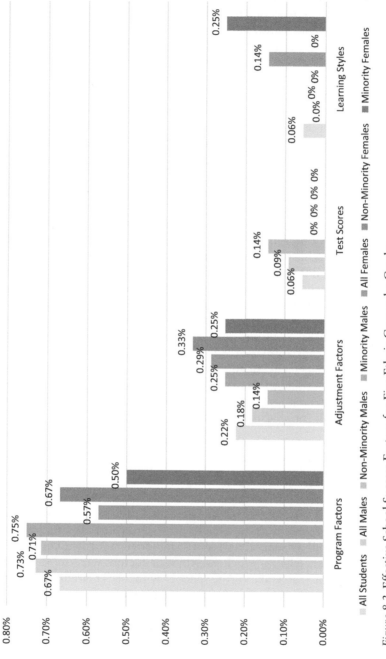

Figure 8.3 Effective School Success Factors for Five Ethnic Groups by Gender

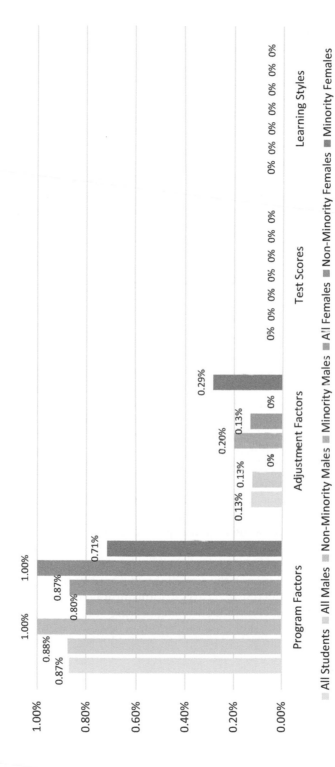

Figure 8.4 Retention Success Factors for Five Ethnic Groups by Gender

while test scores accounted for only 5.6% and learning styles another 5.6%. Internships accounted for 50.0% of the program factors, followed by 25.0% for Summer Bridge, and 25.0% for Freshman Orientation Courses. Thus, students in effective schools utilized academic programs more, but the critical program was clearly Internship Involvement.

For males, 72.7% of the effective school factors described program participation, with 18.2% adjustment factors, and 9.1% test scores. There was little variation between non-minority and minority groups of males. Among females, there was more variation between groups, but program factors still dominated. For the five groups of women, 57.1% of the predictors of effective schools were program factors, followed by 28.6% describing adjustment factors, and 14.3% representing learning styles. Program factors were more frequent among the predictors for non-minority than minority women (66.7% vs. 50%).

Predictors of Freshman-Senior Differences

What factors were more common among seniors compared to freshmen, which might inform retention to graduation in these engineering schools? As shown in Figure 8.4, the long and short answer is program participation. In this conclusion, there were no gender differences and no differences among ethnic groups. For all students, 86.9% of the factors more common among senior students described academic program participation, while only 13.0% described adjustment. The critical programs were Research, Internships, Study Groups, Middle School STEM Programs, Peer Mentoring, Tutoring, and Freshman Orientation Courses, but Research constituted 40.0% of the program factors, and Internships another 20.0%. Thus, 60.0% of the retention factors were research or internship involvement.

In short, making a good adjustment to undergraduate engineering education, through satisfaction with the environment and/or skill in navigating the educational landscape, most facilitates better academic performance. However, matriculating in an effective institution that graduates more students, and survival to the senior year(s), were associated with participation in the academic programs. Principal among these were internship and research participation, although a range of programs made a significant contribution to engineering success. This overall pattern overshadowed ethnic and gender differences.

Conclusion

This analysis of the intersections of gender and ethnicity was undertaken because previous chapters revealed no major gender differences, and certainly no major negative consequences for women engineering students in performance or adjustment analyses. Nonetheless, our focus group studies did reveal female complaints of inequitable treatment, leaving us with the feeling that our picture of success factors might be incomplete. Indeed, the previous literature on women in engineering clearly indicates that women of any description suffer from a poorer adjustment to the overwhelmingly male educational environment. Their adjustment issues include lack of belonging, lack of respect, being ignored, being forced into stereotypical roles, and the prospect of such discomfort continuing throughout their careers. Yet, no matter how embattled women in engineering may be, this is not evident from their academic performance in previous studies as well as in the present study. In no instance did females underperform. Within all ethnic groups except one, their performance was equal to their male counterparts both in test scores and overall GPA. The one exception was Asian females, who exhibited a performance profile that can only be described as over-achievement, for which accolades are due. Their official average GPA of 3.4 far exceeded those of other groups that hover near 3.0. This is consistent with Varma's (2002) finding that Asian women in computer science expressed the most confidence with their math and logic skills. And rather than showing poor adjustment on the multiple measures of adjustment used, most groups of women showed evidence of better adjustment, particularly social adjustment. This confirms our suspicion that women display better social skills, skills that contribute directly to their academic success, and is consistent with the findings of Wao, Lee, and Borman (2010) that women rely more on the social capital of others rather than the institution. Women also reported participating more in academic programs, and these programs appear to be the key factors contributing to their academic success in a series of regression equations.

Among non-minority groups, male or female, White students appear to have the advantage. Their adjustment was better, they participated more in programmatic activities, and most importantly, they had far greater access to internships. Asian students seemed either more inclined to or have more access to opportunities to participate in research. Among minority groups, there were few differences, except that Hispanic females showed evidence of the best adjustment.

As to success factors, although there was variation among the groups, program participation components contributed most to measures of success, comprising on average 62.5% of the success factors. Principal among those were internships and research, although eight program components were represented. Adjustment factors were a distant second and constituted about 25.0% of the success factors. Test scores accounted for only 9.4% of the success factors when they did contribute, with learning styles contributing even less: 3.1%. There were variations in this pattern depending on the measure of success utilized. For GPA, adjustment factors were slightly more important than program participation factors for all students (39.1% vs. 34.8%), but adjustment was more important for males (45.5% vs. 18.2%), although program participation was more important to female success (50% vs. 33.3%), as well as to minority female success (50% vs. 25%). For effective schools and retention, program participation factors constituted 66.7% to 86.9% of the success factors, respectively. Of the eight programs contributing to overall student success, 55.0% were either internships (30.0%) or research (25.0%). Both programs have emerged as important success factors for minorities throughout this research, but White students have far greater access to them than minority students or other non-minority groups.

In sum, this study found no evidence of female disadvantage in engineering. Perhaps the nine measures of adjustment were inadequate to assess the inner feelings of female engineering students, in which case the use of a wider array of measures would be in order, including measures of self-confidence or self-efficacy that have produced gender differences in other studies (e.g., Heyman et al., 2002; Wilson et al., 2015; Vogt, 2003). It also seems clear that focus groups in this and other studies were more effective in revealing uncomfortable gender issues than this survey. With the present measures however, women showed no adjustment disadvantages and Asian and Hispanic women showed far better adjustment than their male counterparts. It may be that women channel their discomfort into hard work; if that is the case, this evidence indicates that they do it successfully.

References

Allendoerfer, C., Wilson, D., Bates, R., Crawford, J., Jones, D., Floyd-Smith, T., Plett, M., Scott, E., & Veilleux, N. (2012). Strategic pathways for success: The influence of outside community on academic engagement. *Journal of Engineering Education, 101*(3), 512–538.

Ayre, M., Mills, J., & Gill, J. (2011). Two steps forward, one step back: Women in professional engineering. *International Journal of Gender, Science and Technology, 3*(2), 293–312.

Billmoria, D., Joy, S., & Liang, X. (2008). Breaking barriers and creating inclusiveness: Lessons of organizational transformation to women faculty in academic science and engineering. *Human Resource Management: Published in Cooperation with the School of Business Administration, The University of Michigan and in alliance with the Society of Human Resources Management, 47*(3), 423–441.

Camacho, M. M., & Lord, S. M. (2013). *The borderlands of education: Latinas in engineering.* Lanhan, MD: Lexington Books.

Cheryan, S., Ziegler, S. A., Montoya, A. K., & Jiang, L. (2017). Why are some STEM fields more gender balanced than others? *Psychological Bulletin, 143*(1), 1–35. doi: http://dx.doi.org/10.1037/bul0000052.ss

Espinosa, L. L. (2008). The academic self-concept of African American and Latina(o) men and women in stem majors. *Journal of Women and Minorities in Science and Engineering, 14*(2), 177–203.

Felder, R. M., Felder, G. N., & Dietz, E. J. (2002). The effects of personality type on engineering student performance and attitudes. *Journal of Engineering Education, 91*(1), 3–17.

Felder, R. M., & Silverman, L. K. (1988). Learning and teaching styles in engineering education. *Engineering Education, 78*(7), 674–681.

Ferreira, M. M. (2002). The research lab: A chilly place for graduate women. *Journal of Women and Minorities in Science and Engineering, 8*(1), 85–98.

Fifolt, M. M., & Abbott, G. (2008). Differential experiences of women and minority engineering students in a cooperative education program. *Journal of Women and Minorities in Science and Engineering, 14*(3), 253–267.

Fleming, J. (1984). *Blacks in college.* San Francisco: Jossey-Bass.

Fleming, J. (2001). The impact of an historically black college on African American students: The case of Lemoyne-Owen college. *Urban Education, 36*(5), 587–610.

Fleming, J. (2012). *Enhancing the academic performance and retention of minorities: What we can learn from program evaluation.* San Francisco: Jossey-Bass.

Gunter, R., & Stambach, A. (2005). Differences in men and women scientists' perceptions of workplace climate. *Journal of Women and Minorities in Science and Engineering, 11*(1), 97–116.

Hanson, S. L. (2006). Insights from vignettes: African American women's perceptions of discrimination in the science classroom. *Journal of Women and Minorities in Science and Engineering, 12*(1), 11–34.

Hardy, F. R. (1974). Social origins of American scientists and scholars. *Science, 185*(4150), 497–586.

Heyman, G. D., Martyna, B., & Bhatia, S. (2002). Gender and achievement-related beliefs among engineering students. *Journal of Women and Minorities in Science and Engineering, 8*(1), 43–54. doi: http://dx. doi.org/10.1615/ JwomenMinorScienEng.v8.i1.30.

Infante-Perea, M., Román-Onsalo, M., & Navarro-Astor, E. (2018). Expected career barriers in building engineering: Does gender matter? *Journal of Women and Minorities in Science and Engineering, 24*(1), 43–49.

Kimball, J., Cole, B., Hobson, M., Watson, K., & Stanley, C. (2008). A study of women engineering students and time to completion of first-year required courses at Texas A&M University. *Journal of Women and Minorities in Science and Engineering, 14*(1), 67–81.

Lee, W. C., Brozina, C., Amelink, C. T., & Jones, B. D. (2017). Motivating incoming engineering students with diverse backgrounds: Assessing a summer bridge program's impact on academic motivation. *Journal of Women and Minorities in Science and Engineering, 23*(2), 121–145.

Litzler, E., & Lorah, J. (2018). Degree Aspirations of undergraduate engineering students at the intersection of race/ethnicity and gender. *Journal of Women and Minorities in Science and Engineering, 24*(2), 165–193.

Litzler, E., Samuelson, C. C., & Lorah, J. A. (2014). Breaking it down: Engineering student STEM confidence at the intersection of race/ethnicity and gender. *Research in Higher Education, 55*(8), 810–832.

Litzler, E., & Young, J. (2012). Understanding the risk of attrition in undergraduate engineering: Results from the project to assess climate in engineering. *Journal of Engineering Education, 101*(2), 319–345.

Marra, R. M., Rodgers, K. A., & Shen, D. (2012). Leaving engineering: A multi-year single institution study. *Journal of Engineering Education, 101*(1), 6–27.

Martin, J. P., Simmons, D. R., & Yu, S. L. (2013). The role of social capital in the experiences of Hispanic women engineering majors. *Journal of Engineering Education, 102*(2), 227–243.

Metz, S. S., Brainard, S., & Gillmore, G. (1999). National WEPAN pilot climate survey exploring the environment for undergraduate engineering students. In *Technology and society, 1999. Women and technology: Historical, societal, and professional perspectives* (pp. 61–72). New York: IEEE.

Micari, M., Pazos, P., & Hartmann, M. J. Z. (2007). A matter of confidence: Gender differences in attitudes toward engaging in lab and course work in undergraduate engineering. *Journal of Women and Minorities in Science and Engineering, 13*(3), 279–293.

Mills, J., Bastalich, W., Franzway, S., Gill, J., & Sharp, R. (2006). Engineering in Australia: An uncomfortable experience for women. *Journal of Women and Minorities in Science and Engineering, 12*(2/3), 135–154.

Miner, K. N., Diaz, I., & Rinn, A. N. (2017). Incivility, psychological distress, and math self-concept among women and students of color in STEM. *Journal of Women and Minorities in Science and Engineering, 23*(3), 211–230.

Moos, R. H. (1979). *Evaluating educational environments: Procedures, measures, finding and policy implications.* San Francisco: Jossey-Bass.

Nasim, A., Roberts, A., Harrell, J. P., & Young, H. (2005). Non-cognitive predictors of academic achievement for African Americans across cultural contexts. *Journal of Negro Education, 74*(4), 344–348.

NCES (National Center for Education Statistics). (2013). *Digest of education statistics 2013* (NCES 2014–2015). Washington, DC: Author. Retrieved from http://nces.ed.gov/programs/coe/index.asp.

Nixon, C. T., & Frost, A. G. (1990). The study habits and attitudes inventory and its implications for students' success. *Psychological Reports, 66*(3c), 1075–1085.

Ong, M., Wright, C., Espinosa, L., & Orfield, G. (2011). Inside the double bind: A synthesis of empirical research on undergraduate and graduate women of color in science, technology, engineering, and mathematics. *Harvard Educational Review, 81*(2), 172–209.

Pearson, Jr. W. (2005). *Beyond small numbers: Voices of African American PhD chemists.* New York: Elsevier Science.

Perna, L., Lundy-Wagner, V., Drezner, N. D., Gasman, M., Yoon, S., Bose, E., & Gary, S. (2009). The contribution of HBCUs to the preparation of African American women for STEM careers: A case study. *Research in Higher Education, 50*(1), 1–23.

Person, D. R., & Fleming, J. (2012). Who will do math, science, engineering and technology? Academic achievement among minority students in seventeen institutions. In J. Fleming, *Enhancing the retention and academic performance of minorities: What we can learn from program evaluation* (pp. 120–134). San Francisco: Jossey-Bass.

Ramseur, H. (1975). *Continuity and change in black identity: A study of black students at an interracial college.* Unpublished doctoral dissertation, Cambridge, MA: Harvard University.

Reyes, M. E. (2011). Unique challenges for women of color in STEM transferring from community colleges to universities. *Harvard Educational Review, 81*(2), 241–263.

Sedlacek, W. E. (1998). Admissions in higher education: Measuring cognitive and non-cognitive variables. In D. J. Wilds & R. Wilson (Eds.), *Minorities in higher education 1997–98* (pp. 47–68). Washington, DC: American Council on Education.

Sedlacek, W. E. (2004). *Beyond the big test: Noncognitive assessment in higher education*. San Francisco: Jossey-Bass.

Tate, E. D., & Linn, M. C. (2005). How does identity shape the experiences of women of color engineering students? *Journal of Science Education and Technology, 14*(5–6), 483–493.

Van Aken, E. M., Watford, B., & Medina-Borja, A. (1999). The use of focus groups for minority engineering program assessment. *Journal of Engineering Education, 88*(3), 333–343.

Varma, R. (2002). Women in information technology: A case study of undergraduate students in a minority-serving institution. *Bulletin of Science, Technology & Society, 22*(4), 274–282.

Vogt, C. (2003). An account of women's progress in engineering: A social cognitive perspective. *Journal of Women and Minorities in Science and Engineering, 9*(3/4), 217–238.

Wao, H. O., Lee, R. S., & Borman, K. M. (2010). Climate for retention to graduation: A mixed methods investigation of student perceptions of engineering departments and programs. *Journal of Women and Minorities in Science and Engineering, 16*(4), 293–317.

Whitten, B. L., Foster, S. R., Duncombe, M. L., Allen, P. E., Heron, P., McCullough, L. M., Shaw, K. A., Taylor, B. A., & Zorn, H. M. (2004). "Like a family": What works to create friendly and respectful student-faculty interactions. *Journal of Women and Minorities in Science and Engineering, 10*(3), 229–242.

Wilson, D. M., Bates, R., Scott, E. P., Painter, S. M., & Shaffer, J. (2015). Differences in self- efficacy among women and minorities in STEM. *Journal of Women and Minorities in Science and Engineering, 21*(1), 27–46.

Appendix 8.1

Summary of Success Factors by Ethnicity and Gender

Group	Program Factors	Adjustment Factors	Test Scores	Learning Styles
All Students	62.5% 40/64	25.0% 16/64	9.4% 6/64	3.1% 2/64
White Males	42.9% (3/7)	42.9% (3/7)	14.3% (1/7)	—
Asian Males	71.4% (5/7)	14.3% (1/7)	14.3% (1/7)	—
Middle Eastern Males	33.3% (1/3)	—	33.3% (1/3)	33.3% (1/3)
Black Males	80% (4/5)	20% (1/5)	—	—
Hispanic Males	50% (4/8)	37.5% (3/8)	12.5% (1/8)	—
Male Average for Five Ethnic Groups	56.7% 17/30	26.7% 8/30	13.3% 4/30	3.3% 1/30
Non-Minority Males	52.9% 9/17	23.5% 4/17	17.6% 3/17	5.9% 1/17
Minority Males	61.5% 8/13	30.8% 4/13	7.7% 1/13	0%
White Females	83.3% (5/6)	16.7% (1/6)	—	—
Asian Females	85.7% (6/7)	14.3% (1/7)	—	—
Middle Eastern Females	50% (3/6)	33.3% (2/6)	16.7% (1/6)	—
Black Females	83.3% (5/6)	16.7% (1/6)	—	—
Hispanic Females	44.4% (4/9)	33.3% (3/9)	11.1% (1/9)	11.1% (1/9)
Female Average for Five Ethnic Groups	67.6% 23/34	23.5% 8/34	5.9% 2/34	2.9% 1/34
Non-Minority Females	73.7% 14/19 (5/8)	21.1% (4/19)	5.3% 1/19	—
Minority Females	60.0% 9/15	26.7% 4/15	6.7% 1/15	—

9 Black Engineering Students in Black and White Colleges

Summary

Previous research has shown that matriculation in Black versus White schools makes a huge difference for Black students in terms of adjustment, performance, mentoring, aspirations and future economic well-being. HBCUs have surfaced in the STEM literature for facilitating the development of Black scientists. This study has also shown that HBCUs offer Black students an independent pathway to success in engineering. The purpose of this analysis was to examine the differences between Black students who attended Black and White engineering schools in performance, program participation, measures of adjustment, and the predictors of three measures of "success." As in previous studies, Blacks in Black engineering schools reported better academic adjustment on several measures, but students in White engineering schools had greater access to scholarship support and undergraduate research opportunities. The success factors, that predicted three measures of "success," were very similar for students in both sets of institutions—with program participation accounting for 66.7% of the predictive factors, and with 53.8% of these were either internship involvement (30.7%) or undergraduate research participation (23.1%). Despite differences in institutional experiences, there were few differences in what facilitates success in engineering.

Introduction

In previous analyses of this study, HBCUs have emerged among the most effective engineering schools and as a unique pathway to engineering success for Black students. This chapter directly examines the performance, adjustment and success factors for Black students in Black and White institutions in the online survey study. Previous studies of Blacks in Black and White colleges have found huge advantages for students in Black schools despite their usually lesser resources. Will this prove to be the case for Blacks in Black and White engineering schools?

Background

A long series of studies documents the better adjustment and outcomes for African American students in HBCUs (Allen, 1992; Constantine, 1995; Kim, 2002, 2004; Kim & Conrad, 2006; Shorette & Palmer, 2015). These include greater gains in cognitive skills (Fleming, 1984), more intellectual challenge, faculty-student contact, and peer interaction (Palmer, Davis, & Maramba, 2010; Seifert, Drummond, & Pascarella, 2006), more positive quality of student-faculty relationships (Cokley, 2000; Palmer & Young, 2009), higher aspirations to doctoral level degrees (Freeman & McDonald, 2004), higher self-concept ratings (Berger & Milem, 2000), fewer racial identity conflicts (Fleming, 2002), and more support for high achievers (Fleming, 2004; Fries-Britt, 2004; Fries-Britt & Turner, 2002). Bridges, Kinzie, Laird, and Kuh (2008) cited a no-one-fails-here attitude, experimental pedagogy, service learning, and strategic freshmen seminars as factors in the Black college advantage, while Irvine and Fenwick (2011) maintained that Black teachers in HBCUs have higher expectations for Black students. Holland, Major, Morganson, and Orvis (2011) found that a Black college was more supportive of capitalization activities promoting academic and career development than students in a predominantly White college, regardless of race of students, and that gender was less of an issue in an HBCU than a PWI.

Studies of adjustment specific to minority students in engineering are still limited, but the adjustment factor may be no less critical for them, and may well be more so because of the hostile climate and resulting ethnic isolation often described in the engineering education literature (Brown, Morning, & Watkins, 2005; Landis, 2005; Miner, Diaz, & Rinn, 2017). Studies have addressed the negative impact of the perception of racism on retention (Brown et al., 2005), more frequent departures due to feelings of not belonging in engineering (Marra, Rodgers, & Shen, 2012), lower ratings of inclusiveness in the engineering environment (Lee, Matusovich, & Brown, 2014), reduced intellectual development (i.e., critical thinking) in White compared to Black engineering schools (Fleming, Garcia, & Morning, 1996), lack of support from instructors beginning in high school—the more gifted the student, the less perceived support (Fleming & Morning, 1998), lower expectancies of success (Lee, Brozina, Amelink, & Jones, 2017), and lower ratings of abilities compared to non-minorities in engineering (Ro & Loya, 2015). Better academic adjustment predicted academic performance for STEM students in Black colleges, while demographic factors such as social class were predictors in White colleges (Fleming & Person, 2012). The greater benefits of attending HBCUs extend to females in STEM majors (Perna et al., 2009), as well as to students in engineering, where HBCUS are among the top producers of minority engineers (Chubin, May, & Babco, 2005; Malcolm-Piquex & Malcolm, 2015; Ransom, 2015; Reichert & Absher, 1997). Certain PWI institutions that graduate large numbers of Black students in engineering have dual-degree relationships with HBCUs, thus providing a hidden Black college advantage (Malcolm-Piquex & Malcolm, 2015). Fleming, Smith, Williams, and Bliss (2013) found that for engineering students in two HBCUs, the benefits included challenging academic programs coupled with very caring faculty. Likewise, Hurtado et al. (2011) found greater faculty support for students in the sciences at HBCUs compared to more selective colleges. Thus, the usually hostile and competitive engineering environment was offset by strong interpersonal support in HBCUs. In short, minorities in engineering appear to face similar conditions of unwelcome as minorities in other disciplines, but HBCUs may facilitate adjustment. However, the number of comparative studies is still limited.

Purpose of the Study

From the online survey of undergraduate engineering students, this set of analyses focuses on Black students in Black and White engineering schools in order to: (1) compare the patterns of program participation and adjustment; (2) compare the factors that facilitate success in each educational environment using three measures of success; and (3) draw conclusions about the pros and cons of each environment for Blacks in engineering.

Method

Table 9.1 describes the study sample. Of the total population of 260 Black students, 225 were African American and 35 were African or Caribbean. Male students numbered 137 (or 52.7%), and female students numbered 123 (or 47.3%) of the group. Students in Black engineering schools numbered 105 (or 40.4%), and those in White schools numbered 155 (or 59.6%). As previously described, an online survey was administered via Survey Monkey to students attending 18 responding institutions. Of these, three were HBCUs.

Table 9.1 Study Population

Group	All	Male	Female	African	African,	GPA	Test Score
	n = 260	52.7%	47.3%	American	Caribbean	3.01	Percentile
		n = 137	n = 123	86.5%	13.5%	SD = 0.51	74.7
				n = 225	n = 35		SD = 20.7
Blacks in Black Schools	40.4% (105)	48.6% (51)	51.4% (54)	92.4% n = 97	7.6% n = 8	3.03 SD = 0.57	71.8% SD = 25.5
Blacks in White Schools	59.6% (155)	55.5% (86)	44.5% (69)	82.6% n = 128	17.4% n = 27	2.99 SD = 0.46	76.7% SD = 16.5

Note. Number of students in parentheses. Students in White schools represented 15 universities; students in Black schools represented 3 universities.

The measures included a Program Participation Scale with subscales for 11 program components, measures of College Adjustment, Study Habits, Academic Adjustment, Social Adjustment, Faculty Interactions, Comfort, Scientific Orientation, Community Identity, and a Learning Styles inventory. The details of instrumentation and analysis are presented in Appendices 1–3 of Chapter 6. The analysis first compared students in Black and White colleges on the measures of program participation, adjustment, and learning styles, and then used regression analysis to determine the most important predictors of GPA, matriculation in schools more effective in graduating minorities, and retention, i.e., survival to the senior year(s).

Results I: Comparing Blacks in Black and White Colleges

Table 9.2 presents the comparative results for students in both sets of educational settings. First, are there background differences between Black students in Black and White engineering schools? There were some. Students in White schools had slightly higher test score percentiles (76.7% vs. 71.8%), but there were no differences in GPA, with or without controls for test scores. There were no differences in parental years of education, or in working outside of school. However, students in White schools had higher levels of scholarship support and were much more likely to be NACME Scholars. Thus, any differences were primarily in access to greater financial support in predominantly White schools.

For program participation, students in Black schools reported greater participation in Freshman Orientation Courses as well as Study Groups, but students in White schools reported greater participation in undergraduate Research. On adjustment measures, students in Black schools scored higher on three scales: Academic Adjustment; Faculty Interactions; and Academic Management. On Learning Styles, students in Black schools scored higher on Active and Sequential Learning Styles, while those in White schools scored higher on Reflective and Global Learning Styles. While students in Black schools may participate more in certain programs, they also showed better adjustment in academic rather than social areas. The particular advantage for students in White schools was their greater access to undergraduate research opportunities. In short, while Black schools confer program and adjustment advantages, White schools confer more opportunity advantages—both financial opportunities and research opportunities.

Some research has reported that HBCUs are particularly supportive of Black males (Bonner, 2014; Fleming, 1984; Flowers, 2014; Palmer & Young, 2009), while other work has found that these schools are particularly supportive of female scientists (Malcolm-Piquex & Malcolm, 2015; Perna et al., 2009; Whitten et al., 2004). Such findings prompted a dissection of the sample by gender and school environment. As shown in Table 9.3, despite no significant differences in test scores, males in Black schools achieved higher grades with or without controls for test scores. This finding stands in contrast to the overall

Table 9.2 Black Students in Black and White Universities: Differences in Program Participation, College Adjustment, and Background Measures, Controlling for Test Score Percentile (n = 260)

STUDY VARIABLE	Blacks in Black Schools n = 105	Blacks in White Schools n = 155	F with Test Score Percentile	Effect Size (Eta)
DEMOGRAPHIC & BACKGROUND VARIABLES:				
GPA	3.03 (0.50)	2.99 (0.41)	ns	na
GPA, With Test Score Percentile	3.04 (0.50)	2.99 (0.41)	ns	na
Test Score Percentile	71.8% (2.01)	76.7% (1.65)	3.55~*	.014
Level of Scholarship Support	2.03 (0.06)	2.21 (0.05)	5.91*	.022
NACME Scholar	8.5% 7/82	36.8% 46/125	Fisher = ***	na
RAW PROGRAM PARTICIPATION PERCENTAGES:				
Study Groups	62.6% 51/97	33.6% 43/128	Fisher = .01	
Research With a Faculty Member	8.2% 8/97	25.0% 32/128	Fisher = .001	
PROGRAM PARTICIPATION SCALES:				
Freshman Orientation Course Participation Scale	3.67 (0.21)	2.94 (0.17)	7.48**	.028
Study Group Participation Scale	3.56 (0.23)	2.52 (0.19)	11.99***	.045
Research Participation Scale	1.47 (0.17)	2.00 (0.14)	6.02*	.023
COLLEGE ADJUSTMENT SCALES:				
Academic Adjustment Scale	60.8 (0.81)	57.6 (0.67)	9.37**	.035
Engineering Faculty Interactions Scale	122.3 (2.15)	116.9 (1.77)	3.73~*	.014
Academic Management Scale	122.6 (2.15)	116.1 (1.76)	5.57*	.021

(Continued)

Table 9.2 (Continued)

STUDY VARIABLE	Blacks in Black Schools n = 105	Blacks in White Schools n = 155	F with Test Score Percentile	Effect Size (Eta)
LEARNING STYLES:				
Active	6.40 (0.23)	5.72 (0.19)	5.23*	.020
Reflective	4.60 (0.23)	5.28 (0.19)	5.23*	.020
Sequential	6.83 (0.19)	6.06 (0.16)	9.49**	.036
Global	4.17 (0.19)	4.94 (0.16)	9.49**	.036

Note. Standard error of the mean in parenthesis. Effect size (partial *Eta* squared) given for significant results. Internship participation was 30.5% in Black schools compared to 38.7% in White colleges; the differences were not significant.
* $p < .05$
** $p < .01$
*** $p < .001$

patterns in previous chapters, which typically showed no differences in GPA after controls for test scores. For program participation, those in Black schools participated more in Freshman Orientation Courses, and Study Groups, and scored higher on Sequential Learning Styles. Males in White schools scored higher only on Global Learning Styles. Surprisingly, for Black males, there were no differences on adjustment measures in the two sets of schools.

Black females in Black schools scored higher on program participation in Summer Bridge and Study Groups, on adjustment measures of Academic Adjustment and Scientific Orientation, and on Active and Sequential Learning Styles (Table 9.4). Those in White schools participated more in undergraduate Research, and scored higher on Reflective and Global Learning Styles.

When the comparisons were split by gender, the results were less dramatic, but what becomes noteworthy is the better academic performance of Black males in Black colleges, better academic adjustment of females in Black colleges, and greater research participation by Black females in White schools. In this analysis, then, any adjustment advantages of HBCUs appear to favor Black females, a finding that may help explain the high production of female scientists in these schools.

The comparative results were also split by class status (i.e., underclassmen vs. upperclassmen), on the assumption that upperclassmen have had more experience and more developmental opportunities. Table 9.5 shows that indeed, more differences were found among upperclassmen.

Table 9.3 Black Males in Black and White Universities: Differences in Program Participation, College Adjustment, and Background Measures, Controlling for Test Score Percentile (n = 137)

STUDY VARIABLE	Black Males in Black Schools n = 51	Black Males in White Schools n = 86	F with Test Score Percentile	Effect Size (Eta)
DEMOGRAPHIC & BACKGROUND VARIABLES:				
GPA	3.09 (0.74)	2.96 (0.41)	ns	na
GPA, With Test Score Percentile	3.13 (0.73)	2.94 (0.56)	4.32*	.031
Test Score Percentile	71.8% (2.01)	76.7% (1.65)	4.46*	.032
NACME Scholar[1]	7.1% 3/42	44.8% 30/67	Fisher = ***	na
RAW PROGRAM PARTICIPATION PERCENTAGES:				
Study Groups	50.9% 26/51	26.7% 23/86	Fisher < .01	na
PROGRAM PARTICIPATION SCALES:				
Freshman Orientation Course Participation Scale	3.72 (0.31)	2.91 (0.23)	4.36*	.031
Study Group Participation Scale	3.45 (0.33)	2.32 (0.25)	7.53**	.053
COLLEGE ADJUSTMENT SCALES:	No Results			
LEARNING STYLES:				
Sequential	7.10 (0.28)	6.28 (0.22)	5.38*	.039
Global	3.89 (0.28)	4.72 (0.22)	5.38*	.039

Note. Standard error of the mean in parenthesis. Effect size (partial *Eta* squared) given for significant results. No differences among males in Black and White schools in social class, level of scholarship support, research participation (11.8% of Black males in Black schools reported research participation vs. 20.9% of Black males in White schools), or internship participation (37.3% of Black males in Black schools reported internships vs. 38.4% of Black males in White schools).
[1] All students were not sure that they were NACME Scholars. Unsure students were removed from the analysis.
* p < .05
** p < .01
*** p < .001

Table 9.4 Black Females in Black and White Universities: Differences in Program Participation, College Adjustment, and Background Measures, Controlling for Test Score Percentile (n = 123)

STUDY VARIABLE	Black Females in Black Schools n = 54	Black Females in White Schools n = 69	F with Test Score Percentile	Effect Size (Eta)
DEMOGRAPHIC & BACKGROUND VARIABLES:				
GPA	2.97 (0.66)	3.04 (0.58)	ns	na
GPA, With Test Score Percentile	3.13 (0.73)	2.94 (0.56)	4.32*	.031
Test Score Percentile	74.0% (2.74)	75.9% (2.42)	ns	na
NACME Scholar	10.0% 4/40	27.6% 16/58	Fisher < .05	na
RAW PROGRAM PARTICIPATION PERCENTAGES:				
Summer Bridge	14.8% 8/54	31.9% 22/69	Fisher < .05	na
Research With a Faculty Member	8.2% 5/54	25.0% 18/69	Fisher < .05	na
PROGRAM PARTICIPATION SCALES:				
Summer Bridge Participation Scale	1.71 (0.27)	2.43 (0.24)	4.11*	.033
Study Group Participation Scale	3.68 (0.33)	2.78 (0.29)	4.12*	.033
Research Participation Scale	1.41 (0.23)	2.12 (0.21)	5.16*	.041
COLLEGE ADJUSTMENT SCALES:				
Academic Adjustment Scale	62.1 (1.07)	57.5 (0.95)	10.46**	.080
Scientific Orientation Scale	69.4 (1.62)	65.1 (1.43)	3.84~*	.031
LEARNING STYLES:				
Active	6.49 (0.32)	5.58 (0.28)	4.51*	.036
Reflective	4.51 (0.32)	5.42 (0.28)	4.51*	.036
Sequential	6.56 (0.26)	5.79 (0.23)	4.65*	.037
Global	4.43 (0.26)	5.20 (0.23)	4.65*	.037

Note. Standard error of the mean in parenthesis. Effect size (partial *Eta* squared) given for significant results. There were no differences in social class or level of scholarship support. Black females in Black schools participated in internships somewhat less than those in White schools but the difference did not reach statistical significance (24.1% vs. 39.1%, p < .10).
* p < .05
** p < .01
*** p < .001

Table 9.5 Differences among Black Underclassmen and Upperclassmen in Black and White Schools, Controlling for Test Scores (n = 260)

STUDY VARIABLE	Blacks in Black Schools	Blacks in White Schools	F with Test Score Percentile	Effect Size (Eta)
Underclassmen	n = 40	n = 52		
PROGRAM PARTICIPATION SCALES:				
Study Group Participation Scale	3.02 (0.31)	1.91 (0.31)	5.69*	.060
Social Adjustment Scale	104.8 (2.60)	96.6 (2.28)	5.40*	.058
Upperclassmen:	n = 65	n = 103		
BACKGROUND VARIABLES:				
GPA	2.99 (0.06)	3.03 (0.05)	ns	na
GPA, Controlling for Test Scores	3.01 (0.06)	3.02 (0.05)	ns	na
Test Score Percentile	71.3 (2.73)	75.3 (2.17)	ns	na
NACME Scholars	12.2% 6/49	45.2% 38/84	Fisher = ***	na
PROGRAM PARTICIPATION SCALES:				
Freshman Orientation Course Participation Scale	3.83 (0.26)	2.98 (0.21)	6.60*	.038
Study Group Participation Scale	3.02 (0.31)	1.91 (0.31)	8.22**	.047
Supplemental Instruction Participation Scale	4.46 (0.29)	3.77 (0.23)	3.56~*	.021
Research Participation Scale	1.58 (0.23)	2.31 (0.18)	6.21*	.036
COLLEGE ADJUSTMENT SCALES:				
Academic Adjustment Scale	61.3 (0.99)	57.6 (0.78)	8.46**	.049
Academic Management Scale	124.8 (2.53)	118.2 (2.01)	4.17*	.025
Scientific Orientation Scale	70.7 (1.39)	66.0 (1.11)	6.98**	.041
LEARNING STYLES:				
Global Learning Styles	4.42 (0.24)	5.12 (0.19)	5.44*	.032
Sequential Learning Styles	6.58 (0.24)	5.88 (0.19)	5.44*	.032

Note. Standard error of the mean in parenthesis. Effect size (partial *Eta* squared) given for significant results. In Black schools 12.3% of upperclassmen reported research participation compared to 31.1% in White schools (*p* < .10); in Black schools 41.5% reported internship participation compared to 48.5% in White schools (ns).
* *p* <. 05
** *p* < .01
*** *p* < .001

Among underclassmen, those in Black schools reported greater participation in Study Groups and scored higher on Social Adjustment. Upperclassmen in Black schools reported more participation in three programs—Freshman Orientation Courses, Study Groups, and Supplemental Instruction; and scored higher on three adjustment measures—Academic Adjustment, Academic Management, and Scientific Orientation—and had higher Sequential Learning Styles. Upperclassmen in White schools reported more Research participation and had higher Global Learning Styles. Again, the differences in program participation and adjustment favor students in Black colleges, and more so the longer their matriculation in them.

Results II: Success Factors for Blacks in Black and White Engineering Schools

The comparative analyses suggested different experiential advantages for Black students in each kind of educational environment. Yet the next question is, are there differences in the factors that facilitate their success? That is, are there institutional differences in what most facilitates better academic performance, what distinguishes matriculation in more effective schools with higher minority graduation rates, and what facilitates survival to the senior year(s)?

Tables 9.6 and 9.7 present the regression analyses for the three success variables for students in Black and White schools, respectively. They show that while some of the specific predictor variables may be different, the overall patterns were quite similar. Collectively, there were 13 variables positively predicting the three success criteria—six for students in Black schools, and seven for those in White schools. Similar to the findings presented in the previous Chapter 8 on gender, 61.5% of these were program participation factors; 66.7% in Black schools, and 57.1% in White schools. Also like the results in Chapter 8, 53.8% of all success factors were internships (30.7%) or research (23.1%), and 87.5% of the program participation factors were internships or research. Adjustment factors constituted 15.4% of the success factors. There were only two—Social Adjustment was the most important predictor of GPA in Black schools, and College Adjustment was a predictor of matriculation in effective schools for students in White schools. Note that level of scholarship support was the best predictor of GPA for students in White schools, with those receiving higher levels of support getting better grades. While social class (level of parental education) was a factor in academic performance in White schools, it made no difference to performance in Black schools.

In sum, while predominantly Black and White engineering schools may offer different advantages to Black students, the factors that facilitate their success in engineering were quite similar. Program participation,

Table 9.6 Regression Analyses of Measures of Program Participation and College Adjustment on Success Variables for Black Students in Black Colleges (n = 104)

Dependent Variables	Independent Variables	t	R/R²	F
GPA:				
	Social Adjustment Scale	2.74**		
	Internship Involvement Scale	2.37*		
	Research Participation Scale	1.99*	R = .437 R² = .191	7.96***
Effective Schools Rankings:				
	Test Score Percentile	6.82***		
	Freshman Orientation Course Participation Scale	3.77***		
	Summer Bridge Participation Scale	–3.38***	R = 731 R² = .534	36.7***
Academic Status; i.e., Freshman-Senior Differences:				
	Internship Involvement Scale	2.14**		
	Auditory Learning Styles	2.14*	R = .287 R² =.082	4.57*

Note. Positive predictors in italics.
* p < .05
** p < .01
*** p < .001

especially participation in internships and undergraduate research, was most closely associated with Black student success.

Discussion

A considerable body of research indicates that Black students adjust and perform better in predominantly Black schools, both at the undergraduate and graduate levels. Other work suggests that the adjustment advantage helps to foster the greater production of Black scientists and engineers. The present study has also found that HBCUs were among the engineering schools with the highest graduation rates for minority

Table 9.7 Regression Analyses of Measures of Program Participation and College Adjustment on Success Variables for Black Students in White Colleges (n = 154)

Dependent Variables	Independent Variables	t	R/R²	F
GPA:				
	Level of Scholarship Support	3.37***		
	Research Participation Scale	3.19**		
	Socio-Economic Status	2.11*		
	Middle School Program Participation Scale	–2.38*	R = .382 R² = .146	6.42***
Effective School Rankings:				
	Internship Involvement Scale	3.69***		
	College Adjustment Scale	2.31*	R = .345 R² = .119	10.03***
Academic Status; i.e., Freshman-Senior Differences:				
	Internship Involvement Scale	4.40***		
	Research Participation Scale	3.81***		
	Test Score Percentile	2.32*	R = .476 R² = .227	14.75***

Note. Positive predictors in italics.
* $p < .05$
** $p < .01$
*** $p < .001$

students, and that matriculation in Black engineering schools in and of itself constituted a significant success factor. The preceding, then, suggests that at the very least, Black engineering students in Black schools, compared to those in White schools, would show evidence of better adjustment. This was indeed the case.

Black students in Black schools showed better academic adjustment, i.e., satisfaction with the academic experience, more positive interactions with faculty, and better academic management, i.e., better navigational and pro-active management skills. Such adjustment advantages were more visible the longer students were in school, that is, among upperclassmen compared to underclassmen. The genders did have somewhat different experiences in Black schools, with Black males reporting greater program participation and females reporting greater adjustment advantages. One notable result, however, was that Black males in Black schools achieved higher grades than males in White schools, even though the test scores of Blacks in White schools were higher. It is worth saying that the adjustment advantages were in the academic arena, rather than the social arena. It is not that engineering students in Black schools were happier or had more friends or a better social life; it was that they were more satisfied with the academic experience, learned more how to handle their academic experiences, and had more positive interactions with their engineering faculty—and in the case of females, stronger scientific leanings. The difference was not, then, a better social life, but a better academic fit.

For Black students in White schools, there were compensations. These students reported higher levels of financial support, were much more likely to be recipients of NACME scholarships, and more likely to be engaged in undergraduate research with a faculty member. Thus, what students in White schools lack in adjustment, they gain in opportunity advantages.

Despite the relative advantages and disadvantages of each educational environment, virtually the same factors were associated with success in engineering for these students regardless of institutional environment. For both sets of students, participation in MEP program activities was the key to success. More than half of the predictors of three measures of success were program participation factors. Furthermore, participation in internships and undergraduate research comprised the great bulk of significant program factors—86.5%. The one wrinkle in this formula was that students in White schools had greater access to undergraduate research opportunities than those in Black schools—8.2% in Black schools, and 25.0% in White schools. However, there were no significant differences in internship participation—30.5% in Black schools compared to 38.7% in White colleges.

While students in Black schools may experience better adjustment to the academic arena, adjustment variables contributed little to engineering success, at least not directly. Only 15.4% of the success factors were adjustment variables. Better social adjustment was the best predictor of GPA in Black schools, while better college adjustment was a significant predictor of effective schools for students in White schools. Making a good adjustment in any environment is undoubtedly a good thing; but in the mathematical world of engineering, immersion in the work of engineering provides the best path to success.

References

Allen, W. R. (1992). The color of success: African American college student outcomes at predominantly white and historically black public colleges and universities. *Harvard Educational Review, 62*(1), 26–44.

Berger, J. B., & Milem, J. F. (2000). Exploring the impact of historically black colleges in promoting the development of undergraduates' self-concept. *Journal of College Student Development, 41*(4), 381–394.

Bonner, F. A. (2014). Academically gifted African American males: Modeling achievement in the historically black colleges and universities and predominantly white institutions context. In F. A. Bonner (Ed.), *Building on resilience: Models and frameworks of black male success across the P-20 pipeline* (pp. 109–124). Sterling, VA: Stylus Publishing.

Bridges, B. K., Kinzie, J., Laird, T. F. N., & Kuh, G. D. (2008). Student engagement and student success at historically Black and Hispanic-serving institutions. In M. Gasman, B. Baez, & C. S. V. Turner (Eds.), *Understanding minority-serving institutions* (pp. 217–236). Albany, NY: State University of New York Press.

Brown, A. R., Morning, C., & Watkins, C. (2005). Influence of African American engineering student perceptions of campus climate on graduation rates. *Journal of Engineering Education, 94*(4), 263–271.

Chubin, D. E., May, G. S., & Babco, E. L. (2005). Diversifying the engineering workforce. *Journal of Engineering Education, 94*(1), 73–86.

Cokley, K. (2000). An investigation of academic self-concept and its relationship to academic achievement in African American college students. *Journal of Black Psychology, 26*(2), 148–164.

Constantine, J. M. (1995). The effect of attending historically Black colleges and universities on future wages of Black students. *Industrial & Labor Relations Review, 48*(3), 531–546.

Fleming, J. (1984). *Blacks in college.* San Francisco: Jossey-Bass.

Fleming, J. (2002). Identity and achievement: Black ideology and the SAT in African American college students. In W. R. Allen, M. B. Spencer, & C. O'Connor (Eds.), *African American education: Race community, inequality and achievement* (pp. 77–92). Burlington, MA: Elsevier Science.

Fleming, J. (2004). The significance of historically Black colleges for high achievers: Correlates of standardized test scores in African American students. In M. C. Brown II & K. Freeman (Eds.), *Black colleges: New perspectives on policy and practice* (pp. 29–52). Westport, CT: Praeger.

Fleming, J., Garcia, N., & Morning, C. (1996). The critical thinking skills of minority engineering students: An exploratory study. *Journal of Negro Education, 64*(4), 437–453.

Fleming, J., & Morning, C. (1998). Correlates of the SAT in minority engineering students: An exploratory study. *Journal of Higher Education, 69*(1), 89–108.

Fleming, J., & Person, D. R. (2012). Who will do math, science, engineering and technology? Academic achievement among minority students in seventeen institutions. In J. Fleming, *Enhancing the retention and academic performance of minorities: What we can learn from program evaluation* (pp. 120–134). San Francisco: Jossey-Bass.

Fleming, L. N., Smith, K. C., Williams, D. G., & Bliss, L. B. (2013). *Engineering identity of Black and Hispanic undergraduates: The impact of minority serving institutions.* Proceedings of American Society for Engineering Education Annual Conference and Exposition (ASEE), Atlanta, GA, pp. 23.510.1–23.510.18.

Flowers, A. M. (2014). Gifted, Black, male, and poor in science, technology, engineering, and mathematics. In F. A. Bonner (Ed.), *Building on resilience: Models and frameworks of Black male success across the P-20 pipeline* (pp. 124–139). Sterling, VA: Stylus Publishing.

Freeman, K., & McDonald, N. (2004). Attracting the best and brightest: College choice influences at Black colleges. In M. C. Brown II & K. Freeman (Eds.), *Black colleges: New perspectives on policy and practice* (pp. 53–64). Westport, CT: Praeger.

Fries-Britt, S. (2004). The challenges and needs of high-achieving Black college students. In M. C. Brown II & K. Freeman (Eds.), *Black colleges: New perspectives on policy and practice* (pp. 161–175). Westport, CT: Praeger.

Fries-Britt, S., & Turner, B. (2002). Uneven stories: The experiences of successful Black collegians at a historically Black and a traditionally White campus. *Review of Higher Education, 25*(3), 315–330.

Holland, J. M., Major, D. A., Morganson, V., & Orvis, K. A. (2011). Capitalizing on opportunity outside the classroom: Exploring supports and barriers to the professional development activities of computer science and engineering majors. *Journal of Women and Minorities in Science and Engineering, 17*(2), 173–192.

Hurtado, S., Eagan, M. K., Tran, M., Newman, C., Chang, M. J., & Velasco, P. (2011). "We do science here": Underrepresented students' interactions with faculty in different college contexts. *Journal of Social Issues, 67*(3), 553–579.

Irvine, J. J., & Fenwick, L. T. (2011). Teachers and teaching for the new millennium: The role of HBCUs. *Journal of Negro Education, 80*(3), 197–208.

Kim, M. M. (2002). Historically Black vs. White institutions: Academic development among Black students. *Review of Higher Education, 25*(4), 385–407.

Kim, M. M. (2004). The experience of African-American students in historically Black institutions. *Thought & Action, Summer,* 107–124.

Kim, M. M., & Conrad, C. F. (2006). The impact of historically Black colleges and universities on the academic success of African American students. *Research in Higher Education, 47*(4), 399–427.

Landis, R. (2005). *Retention by design.* New York: National Action Council for Minorities in Engineering.

Lee, W. C., Brozina, C., Amelink, C. T., & Jones, B. D. (2017). Motivating incoming engineering students with diverse backgrounds: Assessing a summer bridge program's impact on academic motivation. *Journal of Women and Minorities in Science and Engineering, 23*(2), 121–145.

Lee, W. C., Matusovich, H. M., & Brown, P. R. (2014). Measuring underrepresented student perceptions of inclusions within engineering departments and universities. *International Journal of Engineering Education, 30*(1), 150–165.

Malcolm-Piquex, L. E., & Malcolm, S. (2015). African American women and men into engineering: Are some pathways smoother than others? In J. B. Slaughter, Y. Tao, & W. Pearson, Jr. (Eds.), *Changing the face of engineering: The African American experience* (pp. 90–119). Baltimore, MD: Johns Hopkins University Press.

Marra, R. M., Rodgers, K. A., & Shen, D. (2012). Leaving engineering: A multi-year single institution study. *Journal of Engineering Education, 101*(1), 6–27.

Miner, K., Diaz, I., & Rinn, A. (2017). Incivility, psychological distress, and math self-concept among women and students of color in STEM. *Journal of Women and Minorities in Science and Engineering, 23*(3), 211–230.

Palmer, R. T., Davis, R. J., & Maramba, D. C. (2010). Role of an HBCU in supporting academic success for underprepared Black males. *The Negro Educational Review, 61*(1–4), 85–10.

Palmer, R. T., & Young, E. M. (2009). Determined to succeed: Salient factors that foster academic success for academically unprepared Black males at a Black college. *Journal of College Student Retention, 10*(4), 465–482.

Perna, L., Lundy-Wagner, V., Drezner, N. D., Gasman, M., Yoon, S., Bose, E., & Gary, S. (2009). The contribution of HBCUs to the preparation of African American women for STEM careers: A case study. *Research in Higher Education, 50*(1), 1–23.

Ransom, T. (2015). Clarifying the contributions of historically Black colleges and universities in engineering education. In J. B. Slaughter, Y. Tao, & W. Pearson, Jr. (Eds.), *Changing the face of engineering: The African American experience* (pp. 120–148). Baltimore, MD: Johns Hopkins University Press.

Reichert, M., & Absher, M. (1997). Taking another look at educating African American engineers: The importance of undergraduate retention. *Journal of Engineering Education, 86*(3), 241–253.

Ro, H. K., & Loya, K. I. (2015). The effect of gender and race intersectionality on student learning outcomes in engineering. *The Review of Higher Education, 38*(3), 359–396.

Seifert, T. A., Drummond, J., & Pascarella, E. T. (2006). African American students' experiences of good practices: A comparison of institutional type. *Journal of College Student Development, 47*(2), 185–205.

Shorette, C. R., & Palmer, R. T. (2015). Historically Black Colleges and Universities (HBCUs): Critical facilitators of non-cognitive skills for Black males. *Western Journal of Black Studies, 39*(1), 18–29.

Whitten, B. L., Foster, S. R., Duncombe, M. L., Allen, P. E., Heron, P., McCullough, L. M., Shaw, K. A., Taylor, B. A., & Zorn, H. M. (2004). "Like a family": What works to create friendly and respectful student-faculty interactions. *Journal of Women and Minorities in Science and Engineering, 10*(3), 229–242.

10 Success Factors for Minorities in Engineering

Summary and Conclusions

Summary

The National Action Council for Minorities in Engineering (NACME) undertook a study of success factors for minorities in engineering that was conducted in three phases: (1) a study of institutional statistics on minorities and non-minorities in engineering at 26 institutions; (2) a focus group study of 176 students at 11 institutions; and (3) an online survey of 1,145 students at 18 institutions, including 632 minorities and 513 non-minorities, assessing the contributions of program participation, college adjustment, and background to academic performance, effective school rankings, and retention to the senior year(s).

Institutional analysis showed that many of the differences reported in the literature appear due to the significantly lower average test scores of minority students. However, among students with roughly similar test scores, minority student performance rivaled that of non-minority counterparts in math grades, engineering grades, and graduation rates. Similarly, when minority 6-year graduation rates were adjusted for differences in student test scores, the five most effective schools included a variety of institutional types and characteristics—Minority-Serving Institutions (both Black and Hispanic), having a hidden Black college advantage from dual degree programs, hands-on engineering experience provided by a cooperative curriculum, visible leadership on behalf of underrepresented minority students and programs, and the importance of administrator-scholars. In other words, there appear to be multiple pathways available for engineering institutions seeking to be leaders in engineering retention.

Focus group interviews were conducted with involved students active in Minorities in Engineering Programs (MEPs) activities. They gravitated to engineering largely by inclination or family influence, were able to do math if not love it, were often groomed by exposure to STEM programs in secondary school and summer bridge, worked in groups enabling access to help when needed, thrived in student professional organizations, and paid little attention to racism, failure, or frequent setbacks. They desired to graduate, solve problems, build things, and have an impact on the world. Their mini-survey responses were correlated with four measures of success: reported GPA; classification; average minority test scores for the institution; and effective school rankings of the institution. The factor-analytic results showed that "success" was associated with exposure to the work of engineering—as in problem- or project-based courses, research, and industry internships.

The survey study showed the most important factors in success were universal—higher test scores, academic management skills, participation in industry internships, and undergraduate research. Non-minority students participated in support programs when test scores were low, but had far greater access to industry internships—the premier program success factor for all students. African American students participated most in MEP program components that enabled a better adjustment to the scientific milieu; attending an HBCU constituted a success factor in its own right. Hispanic students participated less in MEP programs than Black students, but reported a stronger scientific orientation. Their success depended more on interpersonal relationships of varying kinds that may be less available to Black students. There were few gender differences among African American students, and among Hispanic students, females were more social, outgoing, with identities more strongly wedded to their communities. Undergraduate women in engineering across the ethnicity board performed as well or better than males, and showed no overt signs of disadvantage.

In engineering, there is no substitute for strong academic skill sets, but interest and success can be ignited by exposure to engineering itself—in project-based courses, research, and particularly industry internships.

Introduction

The National Action Council for Minorities in Engineering, an umbrella organization for the distribution of scholarships to underrepresented minority students in undergraduate engineering education, has spearheaded a three-pronged study designed to identify factors enhancing the academic success of such students. For the last four decades, the engineering community has been aware of dwindling enrollments that pose a serious national problem (Gibbons, 2008; Slaughter, 2015; Swail & Chubin, 2007). This situation has led to increased attention to the retention of all students whose dropout rates in engineering, while no higher than students in other disciplines, are too high given their abilities, hovering around 50%. Minority engineering dropout rates are higher still—up to 67%, according to several reports (NACME, 2014a, 2014b; Yoder, 2012), although there is considerable ethnic variation. Underrepresented minorities are the fastest growing segments of the American population; thus, they are potentially a resource for the development of scientific talent. It is in the context of this national drive to develop the engineering pipeline that NACME has lead an effort to increase minority participation in engineering that included the establishment of MEPs at scores of universities (Landis, 2005; Pierre, 2015; Swail & Chubin, 2007). NACME institutions are those with strong MEPs that foster the academic progress of minority students, through staff support and multiple program components. Meeting the current challenges effectively requires an increasingly refined knowledge of factors that influence the development of minority engineers, both at the individual level and the program level.

The NACME study was conceived as a study of success factors for minorities in engineering—possible facilitative program factors, university environment factors, background factors, school adjustment factors, and personality factors. By seeking positive factors that facilitate the movement of minority students through the engineering pipeline, somewhat different questions were asked than might otherwise be the case. The focus was on the positive; that is, what works? What does not work was of less concern for this study, at this time.

Why undertake this study? There is a sparse academic literature that focuses solely on minority engineering students. In most studies, engineering students are lumped together with students in other sciences and rarely disaggregated—yet engineers are said to be different from other scientific types. While scientists are said to be driven by the need to know, engineers are driven by the need to solve problems; i.e., to do, to fix, to make work (Holtzapple & Reece, 2003; Slaton, 2010, p. 205). According to Powel (2006), scientists are concerned with fundamentals; engineers with design. Thus, the exclusive focus on engineering students may lead to insights otherwise obscured by combining engineers with other students in the sciences.

To help fill in some of the gaps in our understanding of what facilitates minority engineering student success, the NACME study was conducted in three basic phases. The first was to gather institutional statistics from NACME partner universities on the grades, test scores, retention, and graduation rates of minority and non-minority students. (The term "non-minority students" is used loosely, since non-minority in this context includes White, Asian, Middle Eastern, and East Indian students.) The goal of Phase 1 was to find out: (1) how *underrepresented* minority students were performing compared to non-minority (not only White) students; and (2) which schools or which minority engineering programs were most effective in retaining/graduating minority students. Phase 2 was designed to gain a better fix on who these students were to design a comprehensive survey. To do this, focus groups were conducted at 11 institutions, where students participated in discussions of their entry into and experiences in engineering, and filled out mini-questionnaires on similar questions. For Phase 3, an online survey of 1,145 minority and non-minority students attending NACME institutions was conducted to learn more about the effects of their program participation and their adjustment in the engineering environment on grades and retention, as well as the distinguishing influences of schools most effective in graduating them.

Part One: Institutional Statistics

Of the 31 NACME group of engineering institutions, institutional statistics were submitted from 26 of them. They graciously provided official data on overall GPA, and GPA in the basic math, science, and engineering courses that constitute the core curriculum, as well as standardized test scores from SAT or ACT. Retention to graduation data was provided for 1-year, 4-year and 6-year cohorts. This data was disaggregated by gender, as well as for African American, Hispanic, and Native American students. The category of non-minority students was not disaggregated. Because of inconsistencies among institutions in how first- and fourth-year data were calculated, 6-year graduation rates proved to be the best indicator across institutions.

Which Engineering Schools Are Best at Graduating Minorities?

Six-year graduation rates are strongly correlated with student test scores. Adjusting the graduation rates for test scores to remove the variance due to student test ability might give a better sense of what schools do with what they have. When such a regression-based procedure was applied to 6-year graduation rates, the top performers were:

1. Georgia Tech—67.1%.
2. University of Texas, El Paso—59.1%.
3. North Carolina A&T—58.4%.
4. Kettering Institute—58.0%.
5. Virginia Tech—56.4%.

Why do these institutions top the list? We can speculate. Georgia Tech's committed leadership has been extolled for many years (Maton, Watkins-Lewis, Beason, & Hrabowski, 2015; Ransom, 2015), and its dual degree programs with several HBCUs are alleged to provide it with a hidden Black college advantage (Malcolm-Piquex & Malcolm, 2015; Sidbury, Johnson, & Burton, 2015). UTEP is a Minority-Serving Institution, and although Hispanic-Serving Institutions are not known for conferring the same educational advantage as HBCUs because their designation is statistical rather than mission-oriented (Bridges, Kinzie, Laird, & Kuh, 2008; Conrad & Gasman, 2015; Contreras, Malcolm, & Bensimon, 2008; Nunez, Hurtado, & Galdeano, 2015), UTEP has distinguished itself as a National Science Foundation Model Institution of Excellence. North Carolina A&T has been among the top producers of African American engineers for many years (Ransom, 2015), as have a number of other HBCUs. Kettering Institute has a unique curriculum with alternating semesters of coursework and industry internships, providing immersion in the work of engineering (Fifolt & Abbott, 2008; Fleming, 2016; see Chapter 5). Virginia Tech has been extolled for its deliberate efforts to pave the way for African Americans to attend this formerly all-White institution (Crichton, 2003), its presidential support for underrepresented minorities (Korth, 2016), and supporting the development of effective retention programs staffed by administrative leaders with noted scholarship records (e.g., Lee, Brozina, Amelink, & Jones, 2017; Lent et al., 2016). In short, there are multiple pathways available for engineering schools seeking to be leaders in minority student retention.

How do Minority Students Perform Compared to Non-Minority Students?

The first difference is that minority student test scores were lower than those of non-minority students (see Figure 3.1), meaning that any comparison would involve students with unequal abilities. So, with test scores controlled, minority students performed lower on overall GPA and GPA in science. However, there were no differences in math GPA or GPA in basic engineering courses. It also seems that in engineering courses (which are of the greatest concern), the performance levels were as equal as could be both before and after controls for test scores. After test scores were controlled, there were no significant differences between minorities and non-minorities in 1-year retention or 4- or 6-year graduation rates. Women get better grades, but only non-minority women also have higher retention rates. This is consistent with other studies showing that women perform best in engineering, a result found across ethnic groups (Lord et al., 2009).

Part Two: Focus Groups

After looking at institutional data, the next step was to meet some of our minority engineering students, to get a feel for them, to make decisions about how to structure the planned survey. As an exploratory method, focus groups originally employed in marketing research, have found their way into educational research in engineering (Fleming, 2012; Kontio, Lehtola, & Bragge, 2004; Van Aken, Watford, & Medina-Borja, 1999). Focus group conversations were arranged with 176 students in groups of 6–12 at 11 of the schools that provided institutional data. The questions fell into nine categories—early success factors, pre-college program experience, summer bridge experience, minority status issues, program participation, faculty issues, job experience, career outlook, and advice to other minority students in engineering. These students were probably not representative of the minority engineering student population. They were involved, known to staff, and inclined to participate in this forum for no compensation save pizza. They were an inspiring group of prob-lem solvers who did not whine or complain as many students do. They were focused on getting the job done, and eagerly awaiting the time when they can make the world a better place, and of course, improve their own financial and career opportunities. The highlights of those conversations follow.

Pathways to Engineering

Are engineers born or made? Are they engineers by natural inclination or can they develop through programmatic efforts? These students talked as if they were born to be engineers, or as though engineering were a natural culmination of their abilities. Program influence was not evident until the high school years. Few of these students were involved in science-related programs in their elementary or middle school years. Most of the pro-gram involvement in high school was in summer bridge programs when they entered engineering school.

About Math

Math is often described as the primary pathway to success in engineering (e.g., Pearson & Miller, 2012), so they were specifically asked about their relationship to mathematics. Some students said that they were very good at math, and beamed with quiet pride. Others admitted to struggling with math because math in engineering is harder than high school math. Still others tried to explain that they can all do it, but it isn't about the math; it's about the problem solving.

The Engineering School Experience

The college experience is an area that usually provokes a lot of emotion from minority students; but engineering students didn't have much com-ment. When asked how they liked their instructors, most students just shrugged and said that some were good, and some weren't. When asked about prejudice and racism, they also had little to say. African American students acknowledged that there was racism, but did not belabor the issue. They saw prejudice as just another problem to solve that made them work harder.

Working Together

So, if these students were not complaining about their instructors or com-plaining about racism, what were they doing? They were working and studying together. They work in groups, within those created by courses or MEP program activities, or in study groups that they create. This appears to be the way that they succeed, and compensate for any instructor difficulties—by helping each other.

Goals

What are their goals? The short version is that they want to graduate from engineering school and go to work. In the main, these students were doers, not researchers, who wanted to use their problem-solving skills in the real world.

Analysis of Focus Group Mini-Surveys

Focus groups can skew results because they may be dominated by strong personalities. So, it is common to administer a short questionnaire so that all participants can voice their opinions (Guiffrida, 2005; Kao & Tienda, 1998; Litchfield & Javernick-Will, 2015). One hundred and forty-four students—comprised of 51.4% African American, 36.8% Hispanic, 11.8% of "Other" ethnicities, and 41.7% women—completed three open-ended questions about how they became interested in engineering, what it takes to be successful in the engineering program, and their advice to incoming minority students. A fourth question asked for their assessment of the effectiveness of seven academic support program components. Student responses were coded for thematic content and then entered into regression equations against four measures of individual and institutional achievement: student GPA; classification (i.e., retention); average institutional SAT score; and institutional graduation rate ranking.

There were 12 variables positively associated with the measures of success. These 12 variables were then submitted to a factor analysis, which resulted in the isolation of six success factors. These were:

1. Factor 1: *Hands-On/Experience-Based Program Components.* These program components, judged more effective by students high-performing by multiple definitions, describe program components that provide practical engineering experience: project or problem-based courses, research experience, and industry internships.
2. Factor 2: *Desire for the Opportunities in Engineering.* This might be called the hunger factor or the drive factor. What appears to be a burning desire to be an engineer meets the quest to seize the better earnings and promise of a better life.
3. Factor 3: *Using All Available Resources.* Both as an assessment of what it takes to succeed and as their best advice, high performers advocated taking advantage of the academic support provided for students.
4. Factor 4: *Dedication vs. Time Management.* This bi-polar factor describes two attitudinal approaches to success that may be

complimentary rather than in opposition. One set of students thinks in terms of dedication and the motivation engineering requires, while the other thinks in terms of the time required and time management skills needed to reach the goal.

5. Factor 5: *MEP Programs.* This factor describes students who realize the benefit of academic support programs in engineering because they provide support for much of what it takes to succeed.

A synopsis of these factors might suggest that success for minorities in engineering depends on: a burning desire to be an engineer and reap its rewards; dedication and time management; hands-on experience with the practical work that engineers do in solving problems, in research and in industry; and the wisdom to use all available resources, including embracing MEP programs.

In sum, from focus group conversations, the vocal and involved student participants suggested that important success factors were skill in math, which overlaps with problem-solving skill, group work, and networking. Their mini-surveys, when examined from the point of view of student "success," highlighted the importance of hands-on experience with engineering as in problem- or project-based courses, research, and internships. These first two phases of the study constituted preparation for the administration of an online survey of the undergraduate engineering experience that would include a more comprehensive focus on college adjustment issues.

Part Three: Survey Study of College Adjustment

The third phase of the NACME "success factors" study administered an online survey to undergraduate engineering students of all ethnicities to conduct a broader assessment of factors that might contribute to minority student success. The study recruited 1,145 students—including 632 minorities, of whom 39.7% were female; and 513 non-minorities, of whom 35.1% were female—attending one of 18 engineering schools where approval for the investigation was granted. There were an additional 41 biracial students who were included only in whole-sample analyses. In addition to background information and student consent for their institutions to provide overall GPA and test scores, the following measures taken from previous studies of student development or evaluations of minority students in STEM programs were completed: (1) MEP Program Participation; (2) College Adjustment; (3) Comfort; (4) Academic Adjustment; (5) Faculty Interactions; (6) Study Habits; (7) Academic Management; (8) Social Adjustment; (9) Scientific Orientation; (10) Community Identity; and (11) Learning Styles. So, in addition to utilizing a formal assessment of MEP Program participation, this phase of the research employed measures of college adjustment and personal orientation to determine their relative importance in accounting for student success with a larger number of students across a larger number of engineering schools.

For each group of students, analysis of covariance was used to determine the association of study measures to each of the dependent variables of GPA, effective school rankings, and academic classification (freshman-senior differences). Regression analyses were then used to select the most important variables contributing to each of the dependent measures of success. Each of the variables selected contributing positive variance was then submitted to a factor analysis to describe unique success factors. An alternate, descriptive method of assessing success factors was featured in Chapter 8. The basic details of instrumentation and method can be found in the Appendices of Chapter 6.

The College Adjustment of Minorities in Engineering

The substantial literature on minority students in college suggests that adjustment to the college environment is a critical issue, that adjustment issues are greater for minority students, and that race/ethnicity usually occasions less friendly treatment, less positive peer interactions, restricted extra-curricular involvement, a less satisfactory social life, and in particular, impoverished interactions with faculty (Ferguson, 2003; Fries-Britt & Turner, 2002; Pascarella & Terenzini, 2005; Rovai, Gallien, & Wighting, 2005; Smith, 2009; Solórzano, Ceja, & Yosso, 2000). The adjustment factor may be no less critical for minorities in engineering, and may well be more so because of the hostile climate and resulting ethnic isolation so often described in the literature (Bonner, 2014; Brown, Morning, & Watkins, 2005; Landis, 2005; Lee, Matusovich, & Brown, 2014). The present study of minority engineering student adjustment suggests that aspects of college adjustment play a role in minority engineering student success, but take a back seat to academic skill sets and MEP program participation.

Enhancing Academic Performance

For the group of 632 minority students, the most important measures positively associated with higher GPA, derived from a regression equation, were:

1. Test scores.
2. Academic Management Scale scores, describing a set of pro-active behaviors employed in negotiating the academic setting, such as schedule management, and a meta-analytic approach to organizing information.
3. Industry Internship Involvement, reported by 25.6% of minority students.
4. Research Participation, reported by only 17.5% of minority engineering students.

Academic skill sets, including tested ability and organizational/managerial skill, appear to come first, followed by two program components that appear again and again throughout this study as critical factors. Research and internships were prominently associated with "success" among focus group participations, and now resurface in a more comprehensive study. Thus, for minority students combined, adjustment issues were less important than academic managerial issues—all of which take a back seat to tested ability.

Effective Schools

In answer to the question of what characteristics distinguish the engineering schools most effective in graduating minority students from others, the most important differences in favor of effective schools were:

1. Internship Involvement.
2. Freshman Orientation Course Participation.
3. Academic Adjustment Scale scores, describing a positive or satisfied adjustment to aspects of the engineering school environment, such as satisfaction with the college environment, effecting a good academic fit, and academic confidence.

Students in effective schools generally reported more involvement in MEP program components, suggesting that their effectiveness is due, in large part, to making greater use of student programming. In this, minority student adjustment to the academic environment was a primary factor in what distinguishes effective schools, but program involvement was even more important.

Freshman-Senior Differences

Given the high dropout rates among minority engineering students, the study investigated "senior factors." Seniors, or upperclassmen, were more often distinguished by reporting greater:

1. Social Adjustment.
2. Research Participation.
3. Industry Internship Involvement.
4. Supplemental Instruction Participation.
5. Study Group Participation.

It appears that survival to the senior years may depend on making a satisfactory social adjustment, as well as greater participation in varied MEP programming. The one variable that appeared in all three regression equations was industry internships, and participation in faculty-guided research appeared in two equations. While research involvement is frequently touted as critical to student success (Kuh, 2008; Lain & Frehill, 2012; Maton & Hrabowski, 2004; Maton, Domingo, Stolle-McAllister, Zimmerman, & Hrabowski, 2009), we find only rare references to the importance of internship participation (Raelin et al., 2014).

Success Factors

When the variables accounting for positive variance in the preceding regression equations were submitted to an analysis to determine common clusters (see Chapter 6), the following four factors emerged:

1. The combination of *Academic and Social Adjustment*, with *Academic Management* (i.e., the organization of talent) loading first.
2. *MEP Program Participation, Type I*, which includes internships and Supplemental Instruction.
3. *MEP Program Participation, Type II*, including research vs. freshman orientation courses, suggesting that different students participate in each kind of program.
4. *MEP Program Participation, Type III*, including study group participation vs. higher test scores, suggesting that more able students less often participate in group study.

What the cluster analysis confirms is that academic skill sets, adjustment, and participation in a variety of program components all make contributions to minority student success in engineering. These factors accounted for 16.8%–31.6% of the variance in dependent variables.

Ethnic Differences

The general literature on minority students in college finds similar issues of adjustment for African American and Hispanic students in the areas of academics, finances, personal/family relationships, and stress management (Chiang, Hunter, & Yeh, 2004; Cole, 2008; Constantine, Wilton, & Caldwell, 2003). The science and engineering literature also finds similar issues of adjustment challenges, more difficult faculty interactions, and teamwork problems for both ethnic groups (Cole & Espinoza, 2008; Dika, 2012; Dika, Pando, & Tempest, 2016; Wolfe, Powell, Schlisserman, & Kirshon, 2016). What similarities or differences did this survey find?

There were 260 Black students, including 225 African Americans and 35 African and Caribbean students, of whom 47.3% were female, and 368 Hispanic students, of whom 34.5% were female. Much like the general literature on minorities in college, this study of success factors has found few ethnic differences between the two groups of students. Black students had higher test score percentiles (74.7% vs. 70.3%), but there were no significant differences in GPA with or without test score controls. Average GPA was 3.03 for the group–2.99 for Black students and 3.05 for Hispanic students. As far as background was concerned, African American students reported higher parental education levels than Hispanic students.

There were, however, several differences worthy of note. African Americans participated more in MEP programs. On program participation measures, there were five differences—and African American students scored higher on all of them, with greater overall Program Participation, Middle School STEM Program Participation, participation in Tutoring, Supplemental Instruction, and Internships. The greatest ethnic divergence was in the critical success factor of internship participation. While 36.9% of African American students reported participation in industry internships, 17.9% of Hispanic students did so. Students who participated in them achieved higher grades and were more often in effective schools, and more often survived to the upperclass years.

On college adjustment measures, there was only one significant difference—Hispanic students reported a stronger Scientific Orientation than African American students. An investigation of the five subscales of this measure indicated that Hispanic students scored significantly higher on three of them—the self-assessment of Math Competence, self-assessment as a Problem-Solver, and belief in a lawful Knowable Universe.

The College Adjustment of Non-Minorities in Engineering

In any comparison of minority and non-minority students, we expect that non-minorities might occupy a more favorable position. Certainly, this is the case in the comparative studies where non-minorities have higher test scores, better grades, higher graduation rates, and a more positive educational experience (Araque, Roldan, & Salguero, 2009; Borrego, Padilla, Zhang, Ohland, & Anderson, 2005; Chen & Weko, 2009; Suresh, 2006/2007). In this study, there were also indications that non-minority engineering students possessed a more positive profile. Their test score percentiles were 16.1% higher (84.3% vs. 72.6%). Their grades were 5.6% higher (3.21 vs. 3.04), but not as high as expected given the disparity in test scores. With test scores controlled, there was little difference in grade averages: 3.16 vs. 3.06. The one other clear advantage owing to non-minorities was their far greater involvement in industry internships than minorities, which surface repeatedly as a significant variable in success (42.5% vs. 25.6%—a factor of 1.7).

However, in all other respects, non-minority engineering students do *not* appear to operate at an advantage. Comparing participation in engineering program offerings, there were significant differences found on eight components, with minority students reporting greater involvement in six—the exceptions being internships and freshman orientation courses. On the assessments of adjustment, there were group differences on five measures, all of them indicating more positive adjustment among minority students to the engineering environment—college adjustment, academic adjustment, study habits, scientific orientation, and community identity. On two other measures which figured prominently in individual and institutional success—academic management skills and faculty interactions—there were no differences between the two groups. In the alleged chilly climate of engineering, there were no differences in student reports of their comfort or social adjustment. We could argue thusly: while there are systemic differences in the entering qualifications of minority engineering students that place them at a disadvantage of roughly 16 percentage-points, this initial disadvantage quickly dissipates as much of the gap in academic performance decreases and their adjustment advantage increases.

Given this state of the profile, we could ask if the adjustment advantages of minority engineering students are not directly related to their greater involvement in program component offerings. Many of the program opportunities available are the direct result of existing or former Minority Engineering Programs established to facilitate the retention of under-represented students through the engineering pipeline to graduation. As such, minority students in the NACME group of institutions are overseen by program directors who encourage program participation among these students, as well as others who might benefit. It does appear that greater program participation aids the success of minority students in that it is associated with higher grades and with retention or longevity in the engineering school pipeline, and in that students in schools more effective in graduating minorities report greater involvement in program activities. The results for this comparative study may well be different than for minorities in other engineering schools, simply because these institutions were selected for the NACME group due to their strong programming for minority students.

In a similar vein, the women in this survey do not appear to operate at the kind of disadvantage that might be expected, given other studies describing their discomfort in the chilly climate of engineering and their sexist treatment owing largely to their numerical imbalance. In this survey, they performed as well or better than male students regardless of ethnicity, while Asian females outperformed all others. Their stalwart profiles appear due to both participation in academic programs and their better social skills.

Conclusion

We draw the following conclusions from this three-phase study of success factors for minorities in engineering:

First, in engineering, ability is paramount. In this case, ability was measured by standardized test scores. No matter how one stands on testing, these scores were important to student success in most cases. That minorities have lower test scores in the institutional study, as well as the student survey, is a problem that should be addressed. Most of the analyses conducted in the institutional and survey studies controlled for the influence of test scores, so that the findings of differences are true if student abilities are roughly equal.

Second, the ability to manage one's ability and the academic environment seems to take a close second to ability in achieving success. Such managerial abilities include the meta-analytic organization of information, the protection of concentration, and the assessment of faculty, as well as effort and time management.

Third, MEP program participation occupies a central position in minority student success. It appears to enable good minority student adjustment, better than that observed for non-minority students. The programs may also offer reasonable substitutes for any lack of faculty attention or guidance.

Fourth, attending an HBCU constitutes a unique pathway to engineering success. There was some evidence that MEP programming in Predominantly White Institutions offers an alternative to what an HBCU provides for its students; i.e., better academic adjustment. This conclusion is consistent with a great deal of prior research, as well as research on students in STEM disciplines, that the HBCU difference is better academic integration rather than social integration (Essien-Wood & Wood, 2013), and strong student programming informed by retention theory (Palmer, Davis, & Peters, 2008; Palmer, Davis, & Thompson, 2010). Regrettably, it should be noted that HSIs do not appear to serve this general function for Hispanic students, although the institutional analyses showed that the University of Texas, El Paso, an HSI, to be uniquely effective in retaining its engineering students.

Fifth, again and again, the study provides evidence that hands-on exposure to the work of engineering—as in problem- or project-based courses, research participation, and particularly industry internships—constitutes the primary success factor.

References

Araque, F., Roldan, C., & Salguero, A. (2009). Factors influencing university dropout rates. *Computers and Education, 53*(3), 563–574.

Borrego, M. J., Padilla, M. A., Zhang, G., Ohland, M. W., & Anderson, T. J. (2005). *Graduation rates, grade-point average, and changes of major of female and minority students entering engineering.* Proceedings, ASEE/IEEE Frontiers in Education Conference, October 19–22, Indianapolis, IN.

Bonner, F. A. (2014). Academically gifted African American males: Modeling achievement in the historically black colleges and universities and predominantly white institutions context. In F. A. Bonner (Ed.), *Building on resilience:*

Models and frameworks of black male success across the P-20 pipeline (pp. 109–124). Sterling, VA: Stylus Publishing.

Bridges, B. K., Kinzie, J., Laird, T. F. N., & Kuh, G. D. (2008). Student engagement and student success at historically Black and Hispanic-serving institutions. In M. Gasman, B. Baez, & C. S. V. Turner (Eds.), *Understanding minority-serving institutions* (pp. 217–236). Albany, NY: State University of New York Press.

Brown, A. R., Morning, C., & Watkins, C. (2005). Influence of African American engineering student perceptions of campus climate on graduation rates. *Journal of Engineering Education, 94*(4), 263–271.

Chen, X., & Weko, T. (2009). *Students who study science, technology, engineering and math (STEM) in postsecondary education*. Washington, DC: U.S. Department of Education.

Chiang, L., Hunter, C. D., & Yeh, C. J. (2004). Coping attitudes, sources, and practices among Black and Latino college students. *Adolescence, 39*(156), 793–815.

Cole, D. (2008). Constructive criticism: The role of student-faculty interactions on African American and Hispanic students' educational gains. *Journal of College Student Development, 49*(6), 587–605. doi: 10.1353/csd.0.0040.

Cole, D., & Espinoza, A. (2008). Examining the academic success of Latino students in science technology engineering and mathematics (STEM) majors. *Journal of College Student Development, 49*(4), 285–300. doi: 10.1353/csd.0.0018.

Conrad, C., & Gasman, M. (2015). *Educating a diverse nation: Lessons from minority-serving institutions*. Cambridge, MA: Harvard University Press.

Constantine, M. G., Wilton, L., & Caldwell, L. D. (2003). The role of social support in moderating the relationship between psychological distress and willingness to seek psychological help among Black and Latino college students. *Journal of College Counseling, 6*(2), 155–165.

Contreras, F. E., Malcolm, L. E., & Bensimon, E. M. (2008). Hispanic-serving institutions: Closeted identity and the production of equitable outcomes for Latino/a students. In M. Gasman, B. Baez, & C. S. V. Turner (Eds.), *Understanding minority-serving institutions* (pp. 71–90). Albany, NY: State University of New York Press.

Crichton, J. (2003). Paving the way: African Americans at Virginia Tech. *Virginia Tech Magazine*, Spring. Retrieved from www.vtmag.vt.edu/spring03/feature2.html.

Dika, S. L. (2012). Relations with faculty as social capital for college students: Evidence from Puerto Rico. *Journal of College Student Development, 53*(4), 596–610.

Dika, S. L., Pando, M. A., & Tempest, B. (2016). *Investigating the role of interaction, attitudes, and intentions for enrollment and persistence in engineering among underrepresented minority students*. Proceedings of the 2016 Annual Conference of the American Society of Engineering Education (ASEE), New Orleans, LA. Retrieved from https://peer.asee.org/17069.

Essien-Wood, I., & Wood, J. L. (2013). Academic and social integration for students of color in STEM: Examining differences between HBCUs and Non-HBCUs. In R. T. Palmer, D. C. Maramba, & M. Gasman (Eds.), *Fostering success of ethnic and racial minorities in STEM: The role of minority serving institutions* (pp. 116–129). New York: Routledge.

Ferguson, R. F. (2003). Teachers' perceptions and expectations and the Black-White test score gap. *Urban Education, 38*(4), 460–507.

Fifolt, M. M., & Abbott, G. (2008). Differential experiences of women and minority engineering students in a cooperative education program. *Journal of Women and Minorities in Science and Engineering, 14*(3), 253–267.

Fleming, J. (2012). Preparing minority students to compete at the cutting edge: The CCNY CREST Center. In J. Fleming, *Enhancing minority students' retention and academic performance: What we can learn from program evaluations* (pp 102–119). San Francisco: Jossey-Bass.

Fleming, J. (2016). *Success factors for minorities in engineering: Analysis of focus group mini-surveys*. Proceedings of the 2016 Annual Conference of the American Society of Engineering Education (ASEE), New Orleans, LA, June 26–29. Retrieved from https://peer.asee.org/16161.

Fries-Britt, S., & Turner, B. (2002). Uneven stories: The experiences of successful Black collegians at a historically Black and a traditionally White campus. *Review of Higher Education, 25*(3), 315–330.

Gibbons, M. T. (2008). *Engineering by the numbers: Engineering degrees in the U.S.* New York: American Society for Engineering Education.

Guiffrida, D. A. (2005). Other mothering as a framework for understanding African American students' definitions of student-centered faculty. *Journal of Higher Education, 76*(6), 701–723.

Holtzapple, M. T., & Reece, W. D. (2003). *Foundations of engineering* (2nd ed.). Boston: McGraw-Hill.

Kao, G., & Tienda, M. (1998). Educational aspirations of minority youth. *American Journal of Education, 106*(3), 349–384.

Kontio, J., Lehtola, L., & Bragge, J. (2004, August). Using the focus group method in software engineering: Obtaining practitioner and user experiences. In *Empirical Software Engineering, 2004. ISESE'04* (pp. 271–280). New York: IEEE.

Korth, R. (2016). VA Tech president calls on university to double minority enrollments. *The Roanoke Times*, November 7.

Kuh, G. (2008). *High-impact educational practices: What they are, who has access to them, and why they matter*. Washington, DC: Association of American Colleges and Universities.

Lain, M. A., & Frehill, L. M. (2012). *2010–2011 NACME graduating scholars survey results*. White Plains, NY: National Action Council for Minorities in Engineering.

Landis, R. (2005). *Retention by design*. New York: National Action Council for Minorities in Engineering.

Lee, W. C., Brozina, C., Amelink, C. T., & Jones, B. D. (2017). Motivating incoming engineering students with diverse backgrounds: Assessing a summer bridge program's impact on academic motivation. *Journal of Women and Minorities in Science and Engineering, 23*(2), 121–145.

Lee, W. C., Matusovich, H. M., & Brown, P. R. (2014). Measuring underrepresented student perceptions of inclusion within engineering departments and universities. *International Journal of Engineering Education, 30*(1), 150–165.

Lent, R. W., Miller, M. J., Smith, P. E., Watford, B. A., Lim, R. H., & Hui, K. (2016). Social cognitive predictors of academic persistence and performance in engineering: Applicability across gender and race/ethnicity. *Journal of Vocational Behavior, 94*(1), 79–88.

Litchfield, K., & Javernick-Will, A. (2015). I am an engineer AND: A mixed methods study of socially engaged engineers. *Journal of Engineering Education, 104*(4), 393–416.

Lord, L. M., Camacho, M. M., Layton, R. A., Long, R. A., Ohland, M. W., & Wasburn, M. H. (2009). Who's persisting in engineering? A comparative analysis of female and male Asian, Black, Hispanic, native American, and White students. *Journal of Women and Minorities in Science and Engineering, 15*(2), 167–190.

Malcolm-Piquex, L. E., & Malcolm, S. (2015). African American women and men into engineering: Are some pathways smoother than others? In J. B. Slaughter, Y. Tao, & W. Pearson, Jr. (Eds.), *Changing the face of engineering: The African American experience* (pp. 90–119). Baltimore, MD: Johns Hopkins University Press.

Maton, K. I., Domingo, M. R. S., Stolle-McAllister, K. E., Zimmerman, J. L., & Hrabowski, F. A. III. (2009). Enhancing the number of African Americans who pursue STEM Ph.Ds.: Meyerhoff scholarship program outcomes, processes, and individual predictors. *Journal of Women and Minorities in Science and Engineering, 15*(1), 15–37.

Maton, K. I., & Hrabowski III, F. A. (2004). Increasing the number of African American PhDs in the sciences and engineering: A strengths-based approach. *American Psychologist, 59*(6), 629–654.

Maton, K. I., Watkins-Lewis, K. M., Beason, T., & Hrabowski, F. A. III. (2015). Enhancing the number of African Americans pursuing the Ph.D. in engineering: Outcomes and processes in the Meyerhoff scholarship program. In J. B. Slaughter, Y. Tao, & W. Pearson, Jr. (Eds.), *Changing the face of engineering: The African American experience* (pp. 354–386). Baltimore, MD: Johns Hopkins University Press.

NACME (National Action Council for Minorities in Engineering). (2014a). African Americans in engineering. *Research & Policy Brief, 4*(1, April), 1–2.

NACME (National Action Council for Minorities in Engineering). (2014b). Latinos in engineering. *Research & Policy Brief, 4*(3, October), 1–2.

Nunez, A. M., Hurtado, S., & Galdeano, E. C. (2015). Why study Hispanic-serving institutions? In A. M. Nunez, S. Hurtado, & E. C. Galdeano (Eds.), *Hispanic-serving institutions: Advancing research and transformative practice* (pp. 1–22). New York: Routledge.

Palmer, R. T., Davis, R. J., & Peters, K. A. (2008). Strategies for increasing African Americans in STEM: A descriptive study of Morgan State University's STEM programs. In N. Gordon (Ed.), *HBCU models of success: Successful models for increasing the pipeline of Black and Hispanic students in STEM areas* (pp. 129–146). New York: Thurgood Marshall College Fund.

Palmer, R. T., Davis, R. J., & Thompson, T. (2010). Theory meets practice: HBCU initiatives that promote academic success among African Americans in STEM. *Journal of College Student Development, 48*(4), 440–443.

Pascarella, E. T., & Terenzini, P. T. (2005). *How college affects students.* San Francisco: Jossey-Bass.

Pearson Jr, W., & Miller, J. D. (2012). Pathways to an engineering career. *Peabody Journal of Education, 87*(1), 46–61.

Pierre, P. A. (2015). A brief history of the collaborative minority engineering effort: A personal account. In J. B. Slaughter, Y. Tao, & W. Pearson, Jr. (Eds.), *Changing the face of engineering: The African American experience* (pp. 13–36). Baltimore, MD: Johns Hopkins University Press.

Powel, J. (2006). Engineering education. In B. A. Osif (Ed.), *Using the engineering literature* (pp. 269–285). New York: Routledge.

Raelin, J. A., Bailey, M. B., Hamann, J., Pendleton, L. K., Reisberg, R., & Whitman, D. L. (2014). The gendered effect of cooperative education, contextual support, and self-efficacy on undergraduate retention. *Journal of Engineering Education, 103*(4), 599–624.

Ransom, T. (2015). Clarifying the contributions of historically black colleges and universities in engineering education. In J. B. Slaughter, Y. Tao, & W. Pearson, Jr. (Eds.), *Changing the face of engineering: The African American experience* (pp. 120–148). Baltimore, MD: Johns Hopkins University Press.

Rovai, A. P., Gallien, L. B., & Wighting, M. J. (2005). Cultural and interpersonal factors affecting African American academic performance in higher education. *Journal of Negro Education, 74*(4), 359–370.

Sidbury, C. K., Johnson, J. S., & Burton, R. Q. (2015). Spelman's dual-degree engineering program: A path for engineering diversification. In J. B. Slaughter, Y. Tao, & W. Pearson, Jr. (Eds.), *Changing the face of engineering: The African American experience* (pp. 335–353). Baltimore, MD: Johns Hopkins University Press.

Slaton, A. (2010). *Race, rigor, and selectivity in U.S. engineering: The history of an occupational color line.* Cambridge, MA: Harvard University Press.

Slaughter, J. B. (2015). Introduction. In J. B. Slaughter, Y. Tao, & W. Pearson, Jr. (Eds.), *Changing the face of engineering: The African American experience* (pp. 1–9). Baltimore, MD: Johns Hopkins University Press.

Smith, W. A. (2009). Campuswide climate: Implications for African American students. In L. C. Tillman (Ed.), *The SAGE handbook of African American education* (pp. 297–309). Thousand Oaks, CA: Sage.

Solórzano, D., Ceja, M., & Yosso, T. (2000). Critical race theory, racial microaggressions, and campus racial climate: The experiences of African American college students. *Journal of Negro Education, 69*(1/2), 60–73.

Suresh, R. (2006/2007). The relationship between barrier courses and persistence in engineering. *Journal of College Student Retention, 8*(2), 215–239.

Swail, W. S., & Chubin, D. E. (2007). *An evaluation of the NACME block grant program.* Virginia Beach, VA: Educational Policy Institute.

Van Aken, E. M., Watford, B., & Medina-Borja, A. (1999). The use of focus groups for minority engineering program assessment. *Journal of Engineering Education, 88*(3), 333–343.

Wolfe, J., Powell, B. A., Schlisserman, S., & Kirshon, A. (2016). *Teamwork in engineering undergraduate classes: What problems do students experience?* Proceedings of the 123rd Annual Meeting of the American Association of Engineering Education (ASEE), New Orleans, LA. ID Number 16447.

Yoder, B. (2012). *Going the distance in engineering education: Best practices and strategies for retaining engineering, engineering technology, and computing students.* Washington, DC: American Association for Engineering Education (ASEE).

11 A Postscript on NACME Scholars

The National Action Council for Minorities in Engineering (NACME) provides about $50,000 a year in student scholarships to each of its Block Grant institutions. Periodic investigations have been made to follow the progress of these scholars (e.g., Lain & Frehill, 2012). However, the engineering institutions have the freedom to package these scholarships in any way they choose, and to decide whether or not the students know the source(s) of their financial aid. The result is that not all recipients are aware that they are NACME Scholars. Therefore, it is more difficult to assess how the financial aid or the identity as NACME Scholars might affect student outcomes. In the survey study of this volume, 144 (or 22.8%) of the 632 minority students were not sure if they were NACME Scholars or not, leaving 488 who were sure and 101 (or 20.7%) of these who said they were NACME Scholars. Of this group, 39.5% were female, 60.5% male, 36.5% African American, 5.9% African-Caribbean, 57.2% Hispanic, and 0.4% Native American. There were no differences in these demographics between the two groups of students. This variable was omitted from the survey analyses to prevent loss of subjects.

The test scores of NACME Scholars were significantly higher than those of non-Scholars (76.9% vs. 71.4%), which may not be surprising for scholarship winners. However, NACME Scholars achieved significantly better grades than non-Scholars even after adjusting for test score differences (3.20 vs. 2.99). Examining the 488 students for other differences between the two groups, adjusting for any differences in test scores, showed that there were only three. Scholars participated more in MEP programs, participated more in internships, and participated much more in undergraduate research. When the prediction of the three study measures of "success" was undertaken, the addition of the Scholar variable made a notable difference. Being a NACME Scholar was the best predictor of GPA, and the best predictor of the difference between freshman and seniors, or retention (or survival to the senior year), although it is possible that the scholarships were more frequently given to upperclassmen. For this sample, the best predictor of matriculation in more effective schools, with higher graduation rates for minorities, was attending an HBCU. From the evidence available, it does seem that NACME Scholars distinguish themselves in important ways—by higher academic performance and greater participation in the critical MEP programs of research and internships. There were several interaction effects of interest. Female NACME Scholars reported more internship participation than males, and Black female NACME Scholars reported the most participation. Black NACME Scholars reported greater comfort in the engineering milieu, and better study habits than non-NACME Scholars; there were no differences among Hispanic recipients and non-recipients on these two measures. This suggests that the scholarship was associated with better adjustment among Black students.

Reference

Lain, M. A., & Frehill, L. M. (2012). *2010–2011 NACME graduating scholars survey results.* White Plains, NY: National Action Council for Minorities in Engineering.

12 Policy Implications

A Call to Action

As a follow-up to the release of the 2008 landmark report, *Confronting the "New" American Dilemma: A Data-Based Look at Diversity* (NACME, 2008), NACME convened the 2011 NACME National Symposium in St. Paul, Minnesota (NACME, 2011). Sponsored by NACME board of directors company, 3M, the theme of the Symposium was *Responses to the NACME Calls to Action.*

The NACME Calls to Action were as follows:

K-12 Education

- Infuse STEM throughout the K-12 curriculum via active, hands-on, project-based learning, and introduce students to STEM careers, starting in preschool with awareness activities.
- Identify and emulate model schools and best practices that show success with underrepresented minorities.
- Refuse to accept stereotypes that women and minorities are not good in STEM disciplines.
- Significantly improve guidance counselors' knowledge of STEM careers and college programs, and have them send the message to students that STEM careers pay in terms of salary, prestige, and challenge.
- Have genuinely high expectations for students of color, and do everything necessary to help the students meet those expectations.

Higher Education

- Remove systemic barriers to underrepresented minorities' participation in college by addressing financial aid and admissions policies.
- Fight back against efforts to create "color blind" institutions by developing and implementing legal programs to admit and retain talented underrepresented minorities.

- Create special programs to support and retain underrepresented minorities as they go through college; programs should make available peer groups, mentors, and role models.
- Make underrepresented minority admissions and retention a primary metric of institutional success.
- Use political capital to push government policies that open doors to postsecondary education for underrepresented minorities.

Government

- Develop a national STEM workforce development policy that stretches from preschool to Ph.D. level.
- Fight back against attempts to make college completely "color blind" in admissions, financial aid, and other areas.
- Develop and pass legislation at the national and state levels that establishes programs to increase underrepresented minority participation in STEM education and careers.
- Adopt policies to totally transform the education system to emphasize active, hands-on, project-based learning, rather than lecture and rote memorization.
- Recognize that recruiting underrepresented minorities into STEM careers is a key strategy for stemming the offshoring of jobs and keeping them on American soil; adopt the same strategy as a way to provide a workforce of American citizens for the defense industry.

Business

- Form partnerships with K-12 schools to promote STEM careers and education to underrepresented minority students, including providing STEM employees to serve as role models and mentors, offering on-site internships to students and teachers, and providing access to the latest equipment and software.
- Use political capital to prompt government to address the "'New' American Dilemma" and institute policies that support the education and career preparation of underrepresented minorities.
- Include recruiting underrepresented minorities as part of workforce development and hiring strategies, and broaden college recruiting efforts to include institutions that traditionally enroll large numbers of underrepresented minorities.
- Make diversity a basic part of company values, and support diversity through policies and systems.
- Recognize the demographic changes in the American population and prepare to develop a workforce made up increasingly of women and people of color.

The research findings from the present study expand upon the recommendations for policy and practice just enumerated. "This is 'action' research at its best for it increases our options for implementation and enlarges the policy space within which interventions become 'better bets' with greater potential return on investments" (Chubin, 2015, p. 399). Taken together, this expanded agenda for research, policy, and practice constitutes the core strategy for maintaining the competitiveness required to resolve the " 'New' American Dilemma."

MEP Programs

This study has shown convincingly that MEP programs constitute the bedrock of success for minorities, and for women as well. They appear to be the reason that minorities in engineering do not suffer the same adjustment disadvantages typical of those in the general college population. Within these program components, minorities can turn to staff for support, and to fellow students for help, and find a haven away from the usually cold atmosphere of engineering. Thus, we should:

- Staff, fund, and recognize MEP programs as critical contributors to minority student success in undergraduate engineering education.
- Design MEP program components based on the outcomes of rigorous "action" research and program evaluation.

Industry Internships

From this research, we can be sure that industry internships should be added to the list of high-impact retention to graduation strategies. In analysis after analysis it became clear that immersing students in the real-world work of engineering could capture student imaginations and propel them to successful conclusions. Problem- and project-based courses provide this kind of exposure, but internships top the list as the premier success factor. Thus, we should:

- Insure access to industry internship experiences for all students.
- Build upon the lessons learned from the NACME Career Center and other examples of promising and best practices in connecting minority engineering students to corporate engineering companies, and corporate engineering companies to minority engineering students, for internship experiences.
- Identify and disseminate best practices in internship experiences via "action" research and program evaluation.

Research Opportunities

It was clear before this study that undergraduate research opportunities constituted a critical retention to graduation strategy, and the present findings only confirmed this again and again:

- Increase opportunities for minority students to participate in undergraduate research.
- Replicate the Meyerhoff Scholarship Program (MSP) and compare findings on different campuses over time (see Maton, Watkins-Lewis, Beason, & Hrabowski, 2015).

HBCUs

It should come as no surprise to anyone following minorities in engineering that HBCUs have been and remain prime producers of minority engineering talent. They provide lessons to other schools as to how to nurture talent, offer better academic adjustment, and provide the same access to internships as do other schools. Thus, we should:

- Offer HBCUs strengthened funding for technology infrastructure and capacity in undergraduate engineering education (see Ransom, 2015).
- Pursue inter-institutional collaboration between HBCU and PWI Schools of Engineering to promote the sharing of best practices.
- Strengthen the research capabilities of these institutions in order to offer their greater access to undergraduate research opportunities.

Gender Balancing

The investigations of gender in this volume have found performance advantages for women in engineering, and no adjustment disadvantages despite their uncomfortable position of being in an extreme minority. Despite sexist treatment, they are stalwart performers whose educations can most be improved by more women in engineering. Thus, we should:

- Staff, fund, and support programs that explicitly target the recruitment of more minority women into undergraduate engineering programs.
- Identify and disseminate best practices for retaining and graduating minority women in undergraduate engineering education.

Test Score Differentials

Minority students in engineering, on the average, have lower test scores than their non-minority counterparts. This, in and of itself, is a factor that detracts from success. The problem is systemic, and requires considered attention. Thus, we should:

- Implement reforms in mathematics and science education, particularly advanced course taking, enhanced teacher quality, heightened graduation requirements, increased student achievement, and reduced disparities in performance among subgroups of students.
- Explore alternative strategies to identifying talent for undergraduate engineering education (see NACME, 2006).
- Teach test-wiseness as a component of all pre-engineering and MEP programs (see McPhail, 1981).

A Final Thought

The NACME study of what works in undergraduate engineering education for minority students expands the knowledge base for all key stakeholders committed to resolving the "new" American dilemma: the condition in which the nation's need for engineers (and scientists) is mounting amid a potentially inadequate pool of skilled individuals to meet the demand. The clear message from this study is that active engagement in the variety of support programs provided by the NACME Partner Institutions plays a large role in engineering success. In fact, the data in this study show that program participation accounts for 67% of the success factors. The policy implications derived from the research findings in this study require the immediate attention of K-12 education, higher education, government, and business. The time for *doing*—for taking *action*—is *now*.

References

Chubin, D. E. (2015). Challenges and opportunities for the next generation. In J. B. Slaughter, Y. Tao, & W. Pearson, Jr. (Eds.), *Changing the face of engineering: The African American experience* (pp. 389–408). Baltimore, MD: Johns Hopkins University Press.

Maton, K. I., Watkins-Lewis, K. M., Beason, T., & Hrabowski, F. A. III. (2015). Enhancing the number of African Americans pursuing the PhD in engineering: Outcomes and processes in the Meyerhoff scholarship program. In J. B. Slaughter, Y. Tao, & W. Pearson, Jr. (Eds.), *Changing the face of engineering: The African American experience* (pp. 354–386). Baltimore, MD: Johns Hopkins University Press.

McPhail, I. P. (1981). Why teach test-wiseness? *Journal of Reading, 25*(1), 32–38.

NACME (National Action Council for Minorities in Engineering). (2006). *Vanguard engineering scholars program: A report on student and institutional participation and progress*. White Plains, NY: National Action Council for Minorities in Engineering.

NACME (National Action Council for Minorities in Engineering). (2008). *Confronting the "new" American dilemma: A data-based look at diversity.* White Plains, NY: National Action Council for Minorities in Engineering.

NACME (National Action Council for Minorities in Engineering). (2011). *Comprehensive symposium report. 2011 NACME National Symposium.* White Plains, NY: National Action Council for Minorities in Engineering.

Ransom, T. (2015). Clarifying the contributions of historically Black colleges and universities in engineering education. In J. B. Slaughter, Y. Tao, & W. Pearson, Jr. (Eds.), *Changing the face of engineering: The African American experience* (pp. 120–148). Baltimore, MD: Johns Hopkins University Press.

Index